Van Sambeek & Van Veen Architecten

Freedom of Organization

edited by Erna van Sambeek & Hans Ibelings

NAi Publishers

Foreword

Every intervention in the built environment is a continuation as well as a disturbance of the existing order. This order is always present, since everything follows its own specific rules or system. Organization and freedom are inextricably linked. It is organization that makes freedom possible. With every project it is necessary to establish a new relationship between freedom and organization. It is only in knowing what you are building upon or breaking with that it is possible to achieve substantive renewal. To strike that balance be-tween continuity and interference it is important to consider the place assumed by a project, not just in the literal sense but also in metaphorical senses. • The freedom of organization is the guiding principle for my architecture and this is reflected in this book about the work of Van Sambeek & Van Veen. Like my projects, the book is based on an organizational structure that allows the freedom to read and interpret the work in various ways. • The book is sub-divided by theme, using selected projects to address the seven forms of organization that feature in my work: organization by the absence of something dominant (neutrality), organization by means of exception (exclusivity), by similarity and difference (family), by repetition and sequence (seriality), by the place that something occupies (position), by questioning where something belongs (interiority), and organization by relations to what was, is and will be (time). • Two or more of these seven themes play a role in each project, and it is their combination that imbues architecture and urban planning with richness and depth. One and the same project is revisited in at least two chapters of this book. Each chapter presents the projects in text and image based on one particular aspect. A reader who wants to uncover all the aspects of a specific project can navigate the book in various ways via the contents pages. Each chapter can be understood as an independently complete entity, while the seven chapters together form an overview of an oeuvre. • Erna van Sambeek

Thematic index

Project index

Family

In everyday use, the word 'family' refers to kinship. Similitude and difference are inextricably linked in every family. For me the family is a means of achieving cohesion. By organizing similarities and differences in my architecture it is possible to generate relationships in various ways between related but non-identical components.

Family

Bridges
Family

It is immediately evident that the bridges in Amsterdam's IJburg expansion project belong together, owing to the similarity in form and material: brick abutments and steel lattice trusses in an identical arch form, or a segment of it. This means that all the bridges traverse the water with a comparable 'leap', no matter what distance they span. If the water is wider than the span of the arch then the approaches are extended, while if the expanse of water is narrower than the span then the lattice truss structure is reduced to a smaller segment of the arch. Another common feature is that the parapets and bridge decks cross each other repeatedly, extending and integrating with the surroundings: the embankments, landing stages or a park.

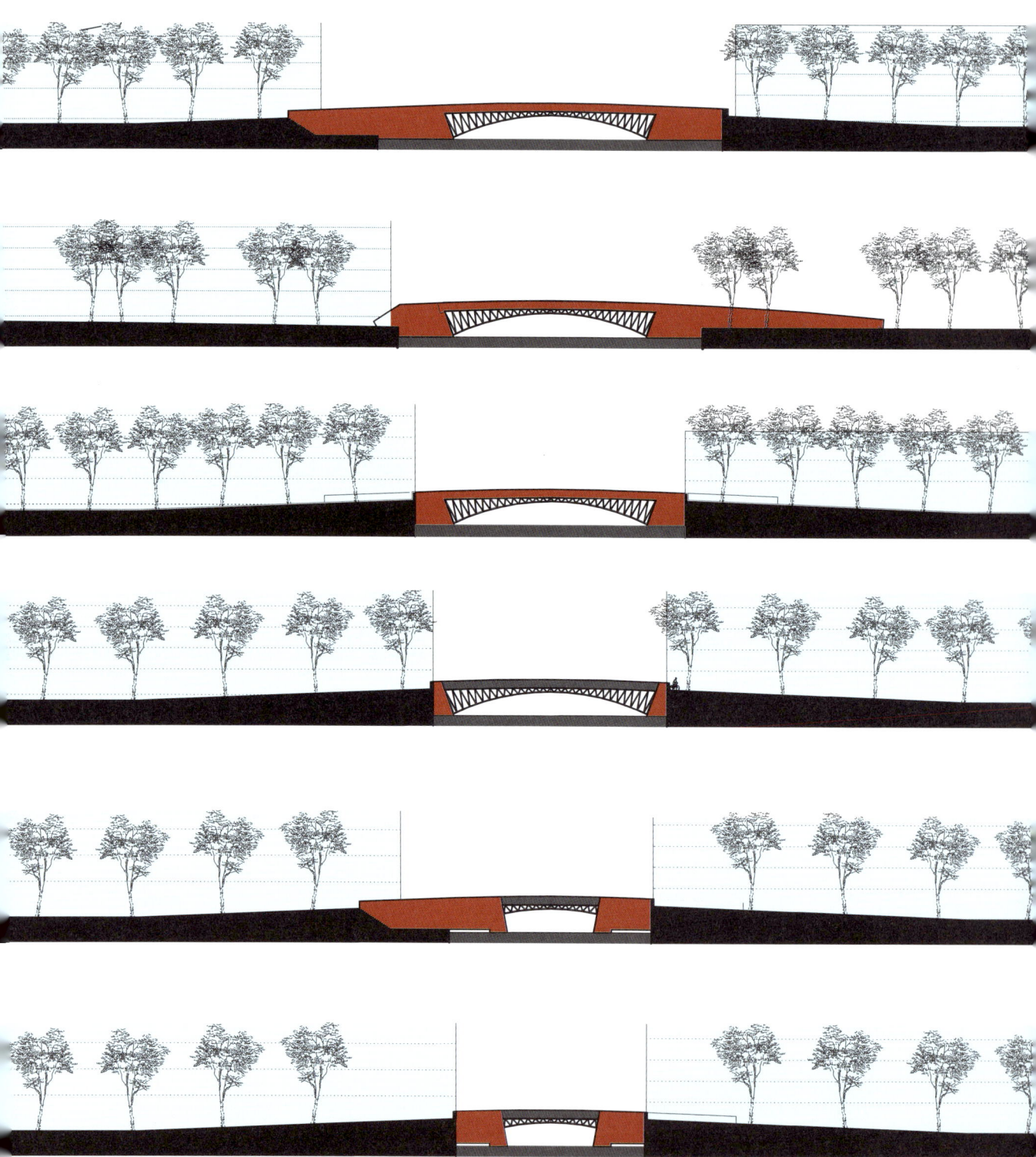

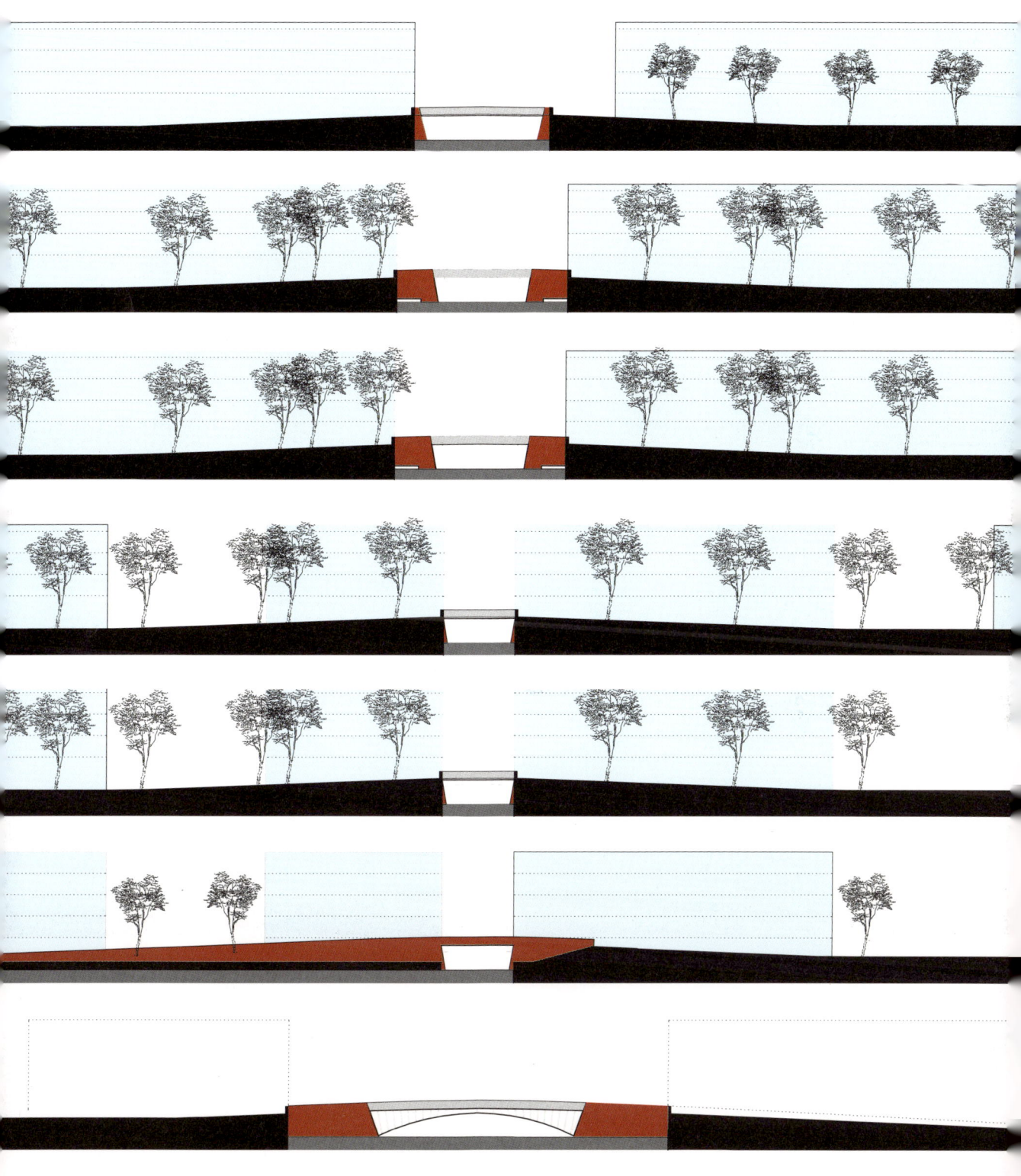

Erasmusgaarde
Family

All the residential buildings in Erasmusgaarde, part of the Zuidwest redevelopment project in The Hague, are of equal height (with the exception of the tower). The dwellings have various storey heights and widths of 6.00, 4.80 or 4.20 metres. This creates 'ladders' within the standard total building height of 13 metres, with varying widths and gaps between the 'rungs'.

19

Tunisia
Family

New houses were designed for various locations in Tunisia for victims of the floods of 1969. In Zaâfrana, new stone houses were added to the scattered clusters of clay houses that were built alongside the unmetalled main road after the floods. Pre-existing clusters were expanded and new groups created. All the stone houses consist of a rectangular room with a flat roof and, alongside it, a partially walled–in space that can be covered over. The houses always have the same core, but are arranged in a variety of combinations and configurations.

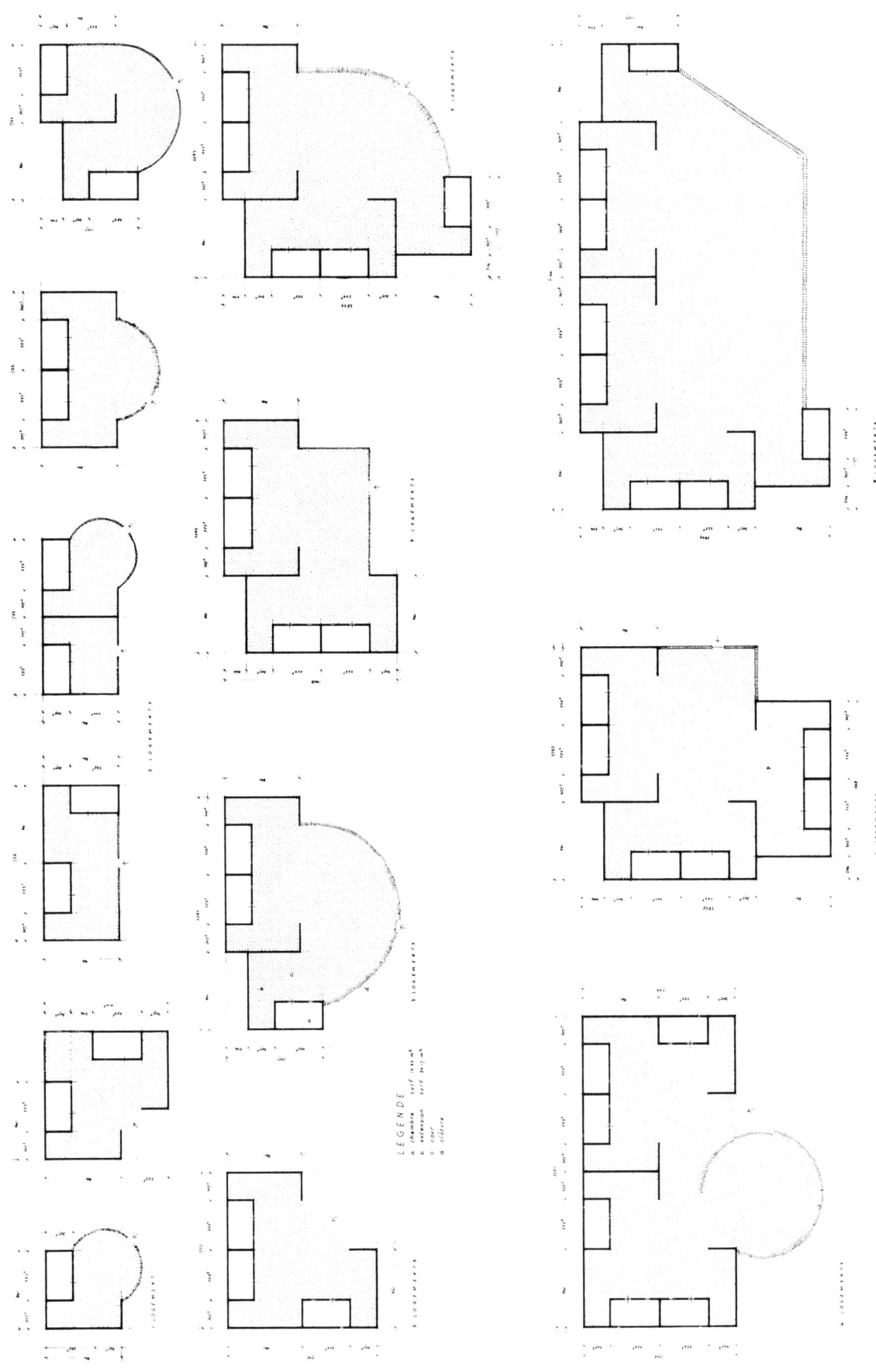

GROUPEMENTS DES LOGEMENTS TYPE ZAAFRANA - 3   SELON LA STRUCTURE FAMILIALE

LÉGENDE
a chambre
b extension
c cour
d clôture

Chassé 100 x 100 Family

family • position • time • interiority • **seriality** 268 • exclusivity • neutrality

In the first design for the '100 x 100' residential complex (the working title that reflected its size in metres) in Breda's Chassé Park the dwellings are members of a spatial family. All the dwellings are of the same type, but subdivided into twelve variants as a result of the different configurations of the same programme within a similar footprint, and the patio is always set in the same place. Together the dwellings form a building, and not merely units that are strung together. Despite its limited height, the size of this building means that it measures up to the other buildings in the park.

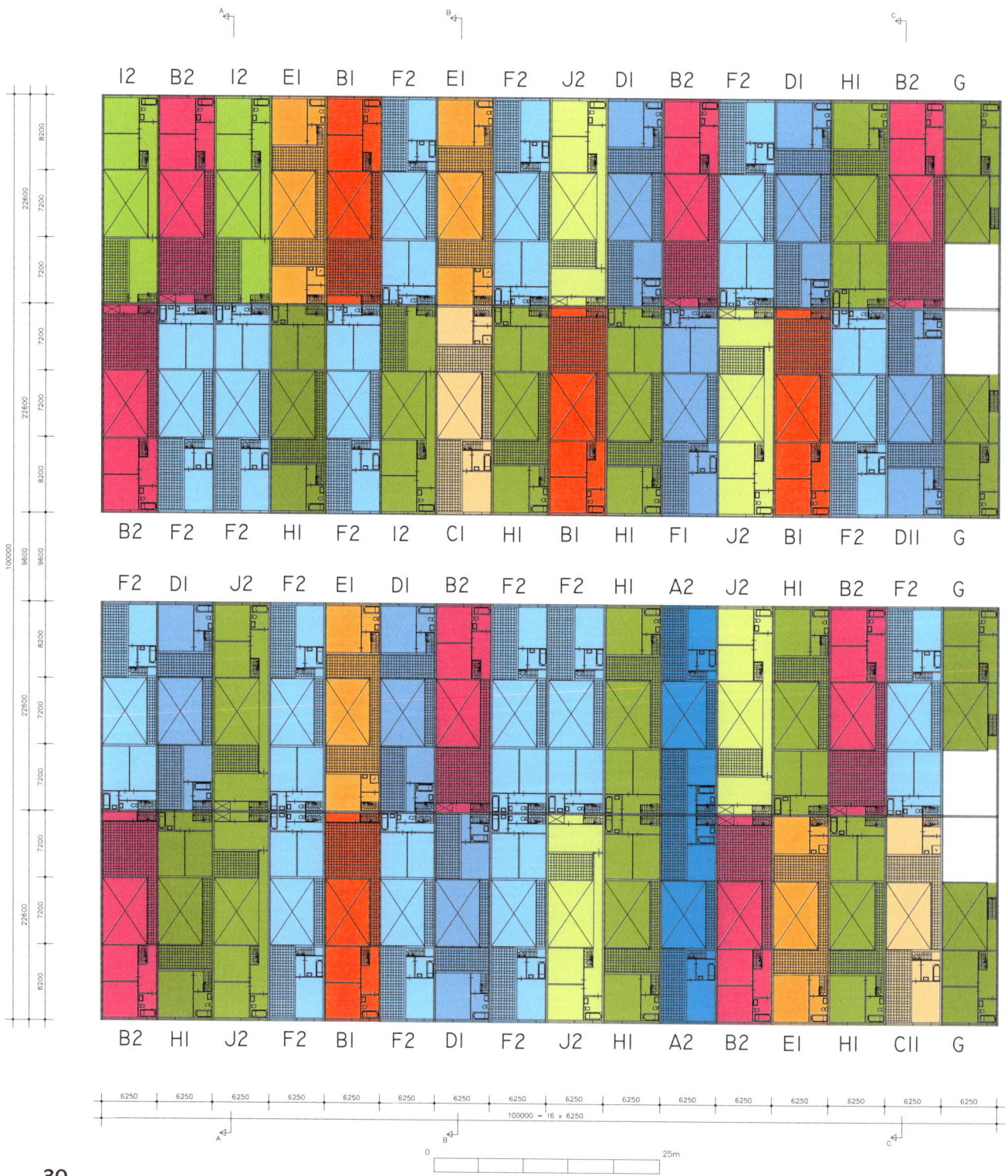

30

A2

B1

B2

C1

D1

E1

F1

F2

G

H1

I1

J2

Witbrant
Family

The patio houses within the urban plan by Jacq. de Brouwer for the Witbrant–Oost neighbourhood in Tilburg form a family. They have a succession and alternation of indoor and outdoor spaces with various gradations of privacy. Sometimes 'living' is rendered expressively on the outside. The houses also vary in width and depth: sometimes they are compact in form, sometimes expansive.

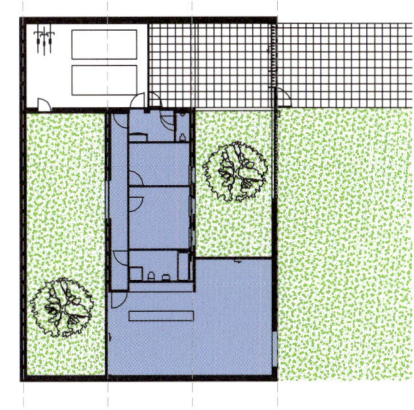

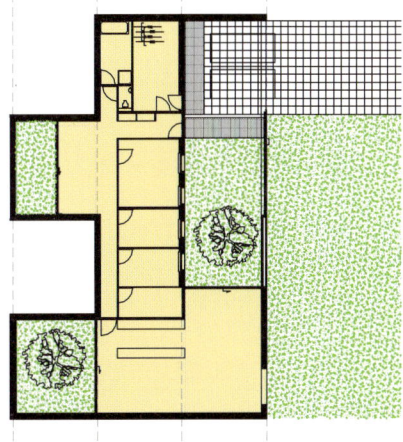

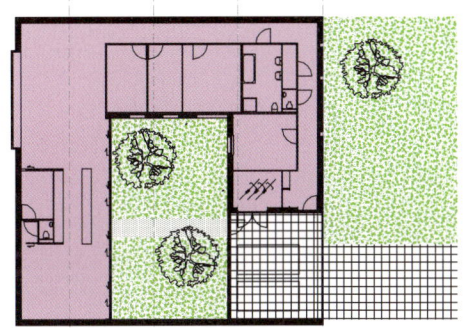

Position

Position refers to the placement of architecture and the stance that architecture assumes in relation to place. The crucial question here is where it belongs, what it corresponds with, and what it turns away from. Sometimes position has a literal bearing on location, but more often on a metaphorical architectural configuration. In the same way that someone is aware of the position that she or he occupies in human company, architecture can also take into account the context where it is set. A building can be understood as an individual in the great company constituted by the city.

Position

Block 34
Position

Block 34 is part of the grid of Amsterdam's ambitious urban expansion district, IJburg. The position of the block within this grid is unusual, since it is surrounded by water and road in equal measure. Being hemmed in between the water and the IJburglaan road-way it is only 23 metres wide, narrower than the grid's normal blocks. • By contrast, for the narrow strip of Block 34, the intended programme was too limited to build on the perimeter. By proceeding from a massive volume of maximum size the block estab-lishes a presence, and rather than being a fragment of a larger-scale block it forms a corner in the grid. It also proved possible to accommodate a more extensive programme. • The four different heights of the block are based on the prescribed percentages for high– and low–rise construction in the build-ing regulations for IJburg. Each of the four sections has a different composition and typology: in the tall flank there are three apartments per storey, while in the low flank there are live-work dwellings and two luxury apartments; in the tall middle section there is one apartment per storey, while in the low middle section there are adaptable two–storey dwellings with flexible floor plans.

42

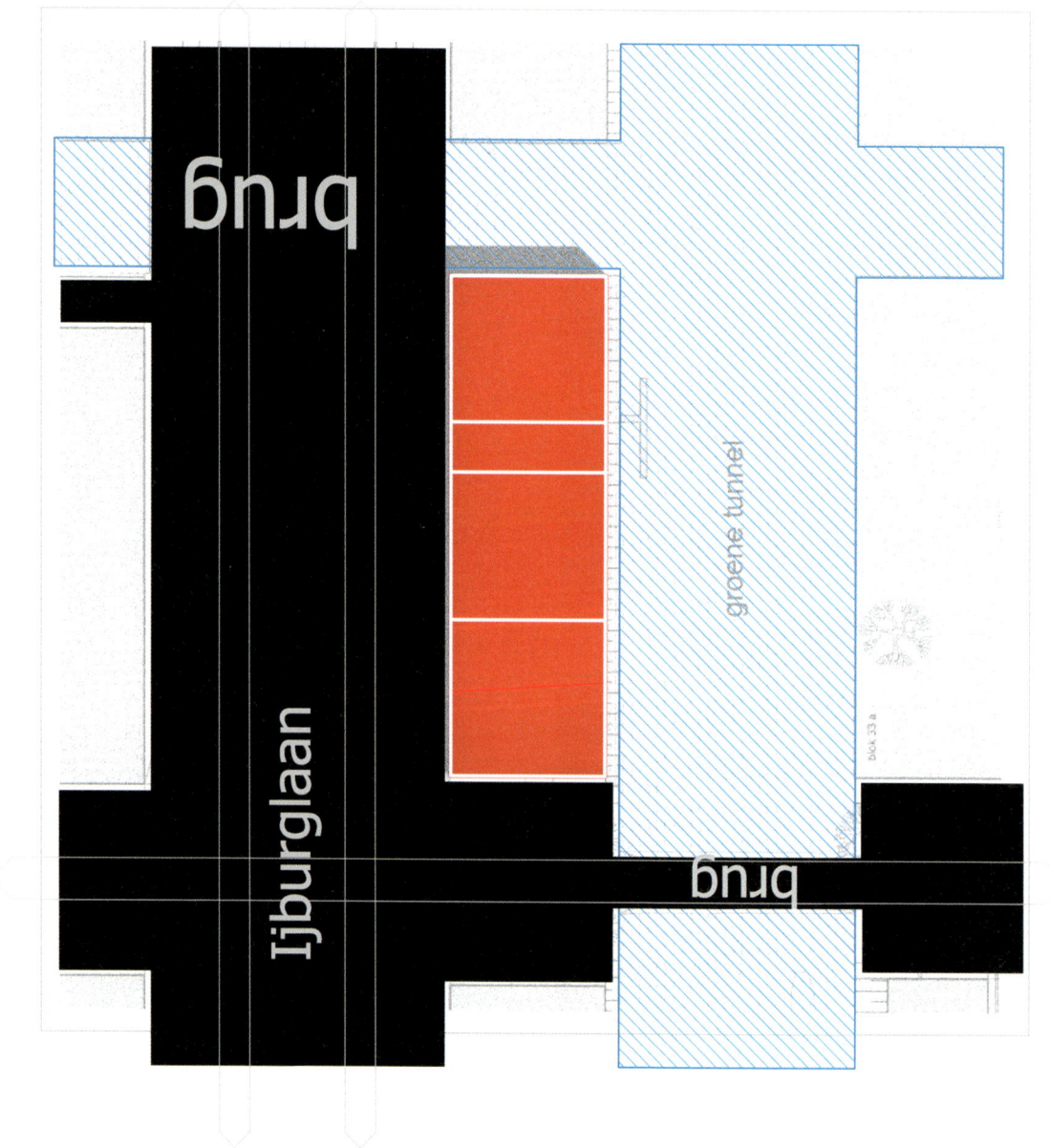

brug

brug

Ijburglaan

groene tunnel

brug

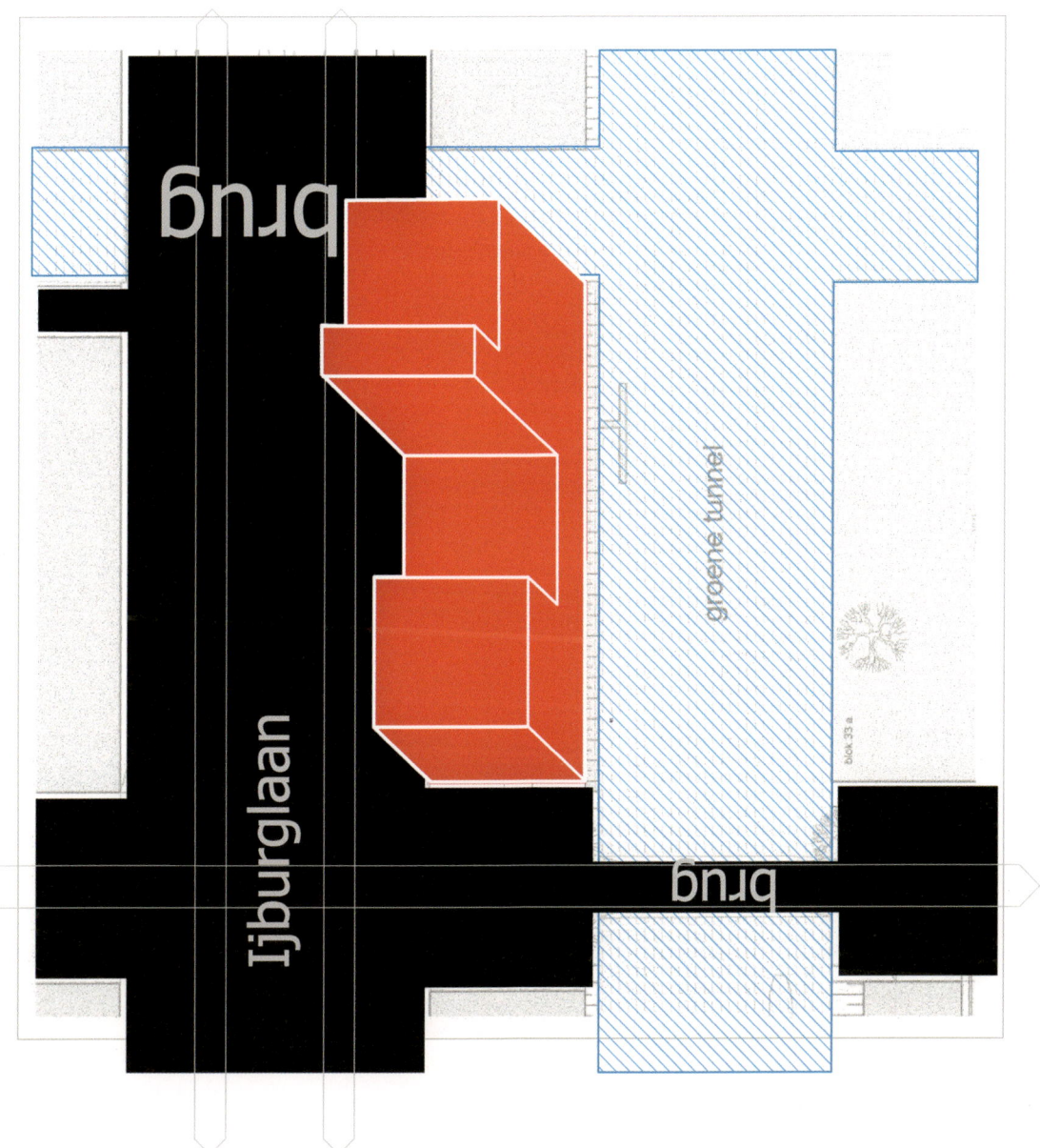

brug

brug

IJburglaan

groene tunnel

blok 33 a

brug

Ijbur

aan

brug

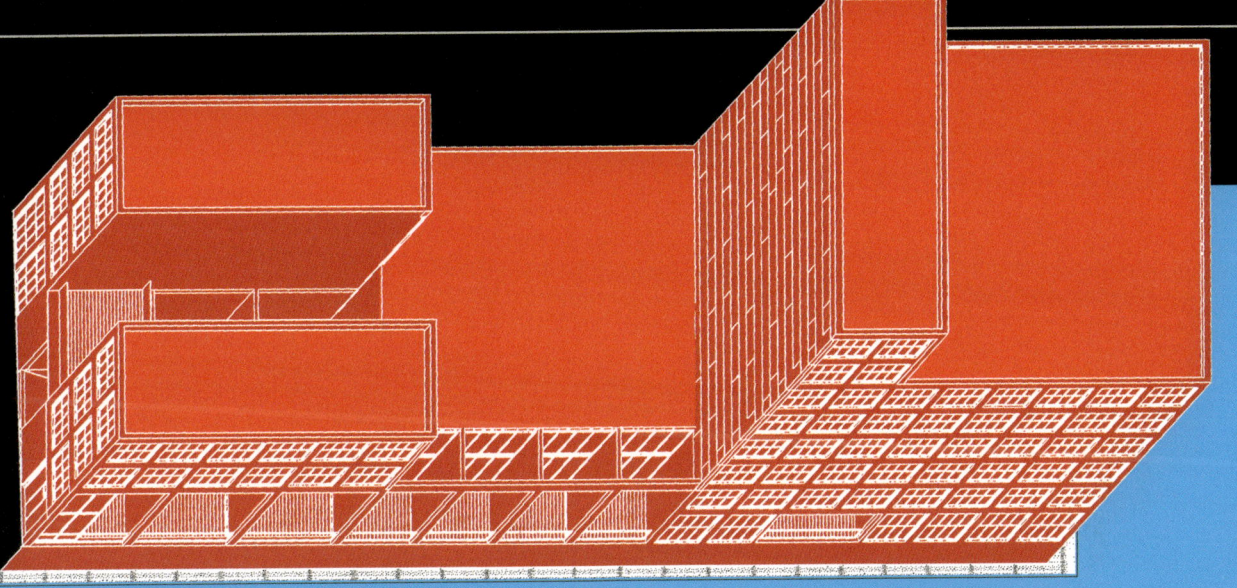

blok 33 a

glaan

brug

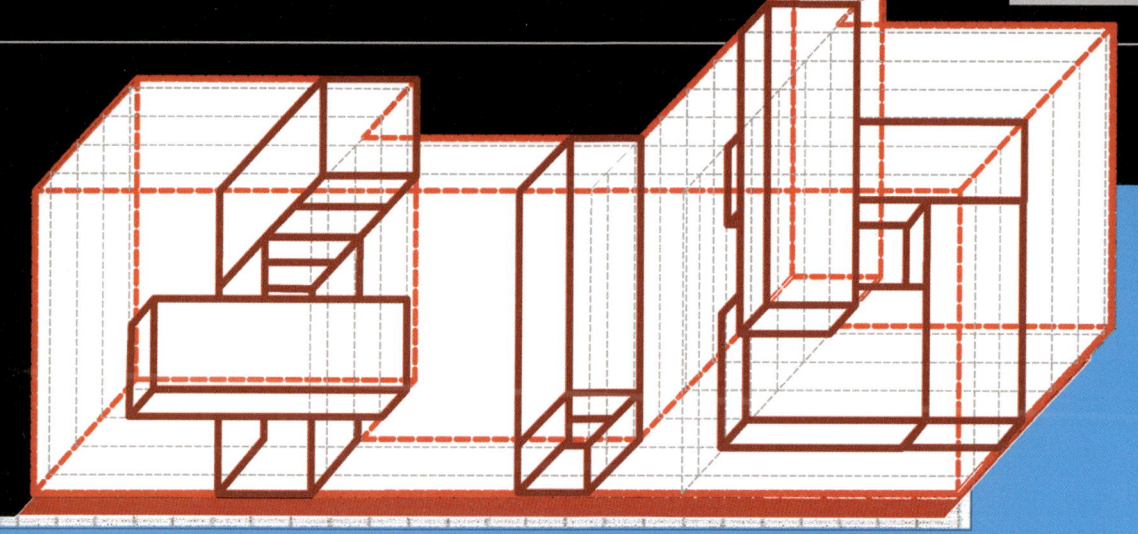

blok 33 a

Waterfront Position

This residential building stands on the edge of a newly built neighbourhood in Bergen op Zoom, overlooking the Westerschelde estuary. It is an autonomous component of a strip that was designed by three different architects. Instead of the prescribed frontal building overlooking the water, a stretched building was designed with an emphatic segmentation, almost dissolving towards to the end. The dwellings have a maximum orientation towards the water.

Oosterpark
Position

The street frontage opposite the Oosterpark in Amsterdam is composed of a succession of single and double properties. The project was conceived as a pair of 'twins' that constitute part of the series of singles and pairs in this elevation. The characteristics of the street frontage are absorbed into the facade's structure, making the project a harmonious component without it being swallowed up invisibly.

Stramanweg
Position

These three blocks stand in a series of five on a trunk road in the Amsterdam borough of Zuid-oost. Each of the blocks is structured by the circulation space, which differs for each storey. The floor plans fold themselves around this figure, so that each dwelling is accessible from the rear as well as via the front entrance. Each of the rear entrances is combined with a balcony on the rear elevation.

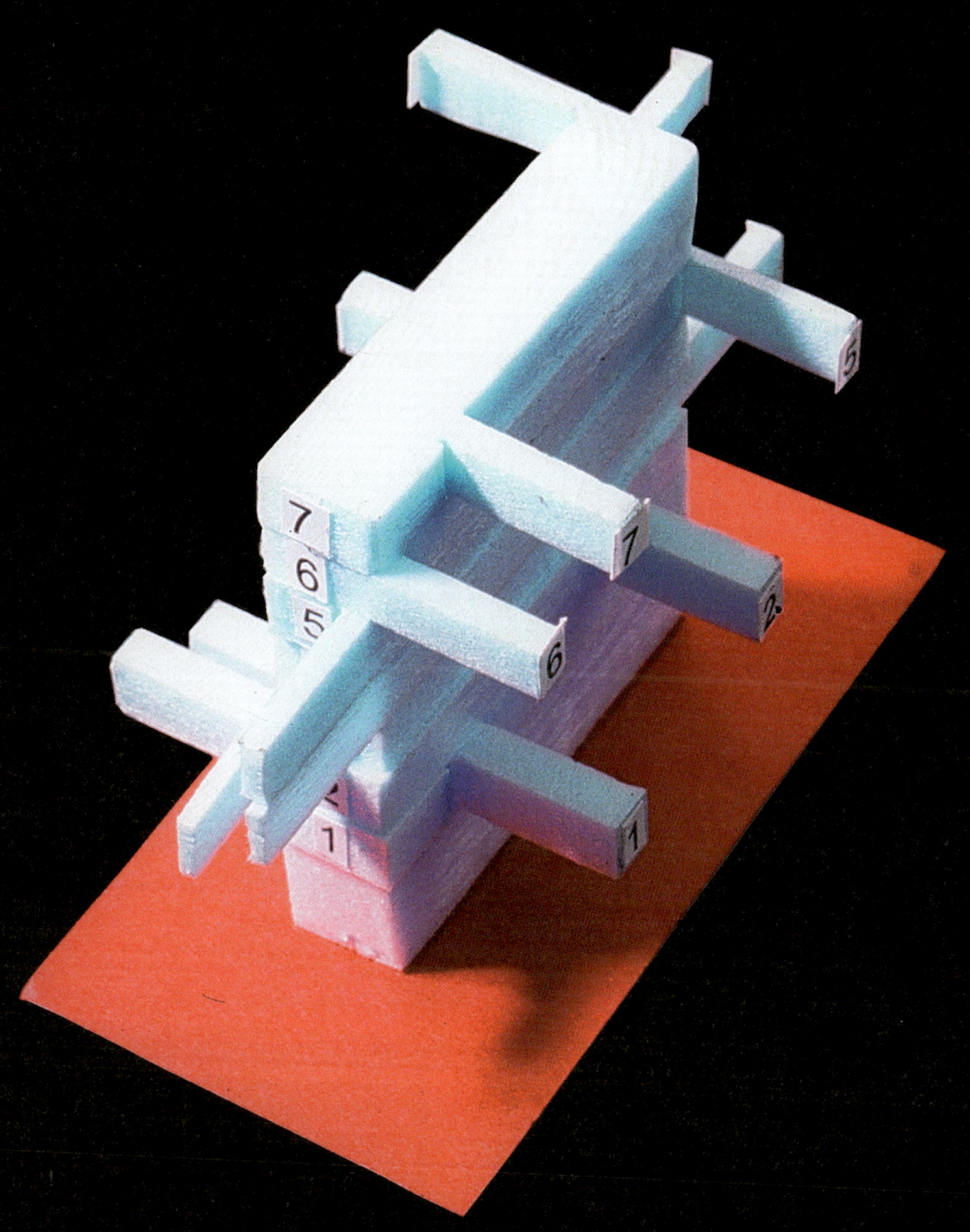

2e verdieping

5e verdieping

1e verdieping

4e verdieping

begane grond

3e verdieping

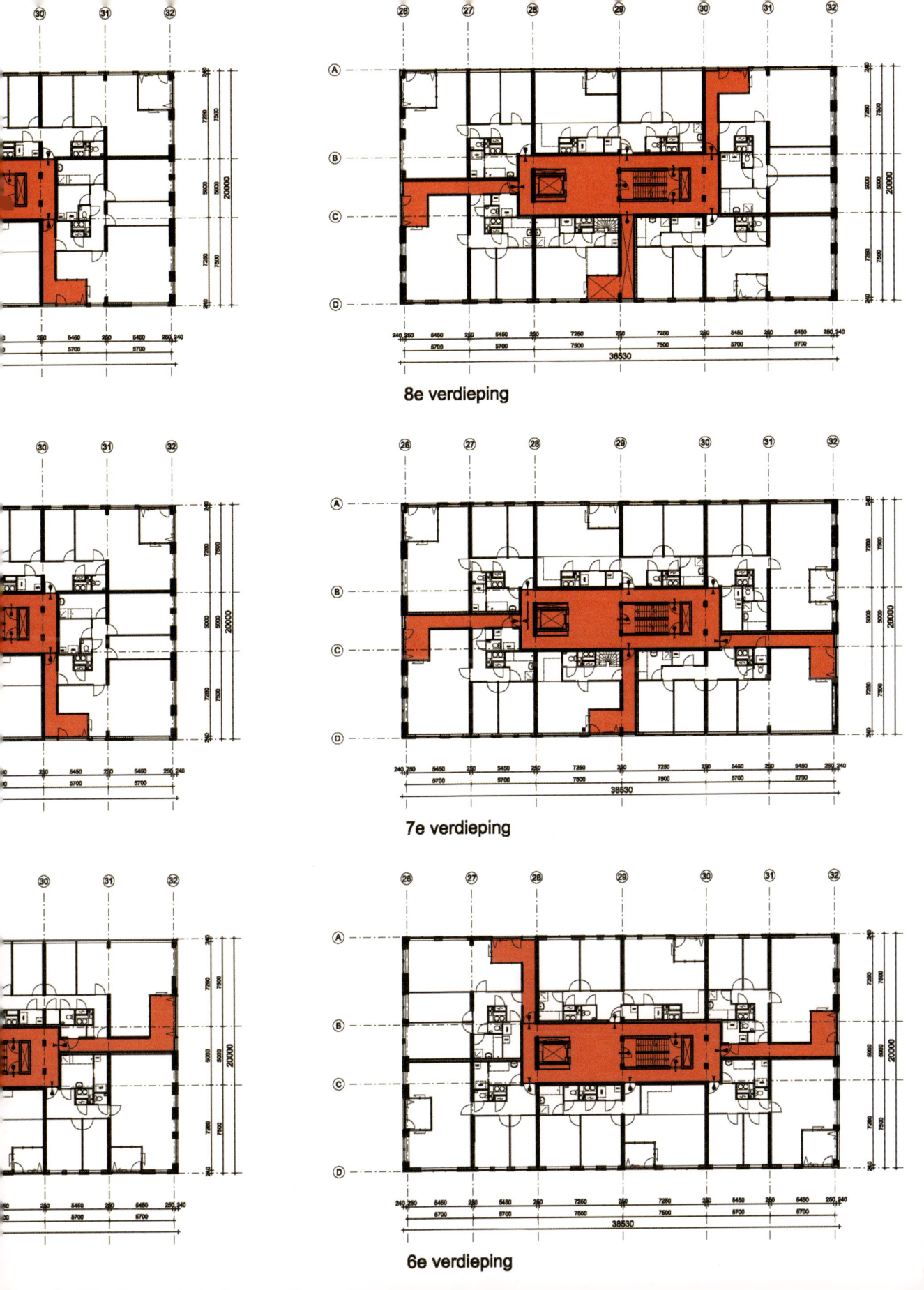

8e verdieping

7e verdieping

6e verdieping

Almere 6A
Position

The name of the '6A' project, a residential block in the city centre plan for Almere Stad (for which OMA created the urban design), indicates where it belongs: with Block 6. Blocks 6 and 6A form an entity. The buildings are of similar height and width and their substructures share the same programme, with underground parking space and a supermarket and shops at street level. Yet 6A also stands there independently, dissociating itself and turning away towards the water. In terms of the programme it also exists independently. In Block 6A, there is a residential programme above the shops, a complete contrast in scale and function to the cinema in Block 6 and the theatre standing opposite. This stark programmatic contrast prompted the covering of the facades of 6A with as much shuttering as possible, in order to make it possible to insulate the dwellings from the nightlife in the evening, especially in the narrow space between the residential block and the cinema. The shutters lend a unity to Block 6A, so that it asserts its presence between the two big buildings.

6

6A

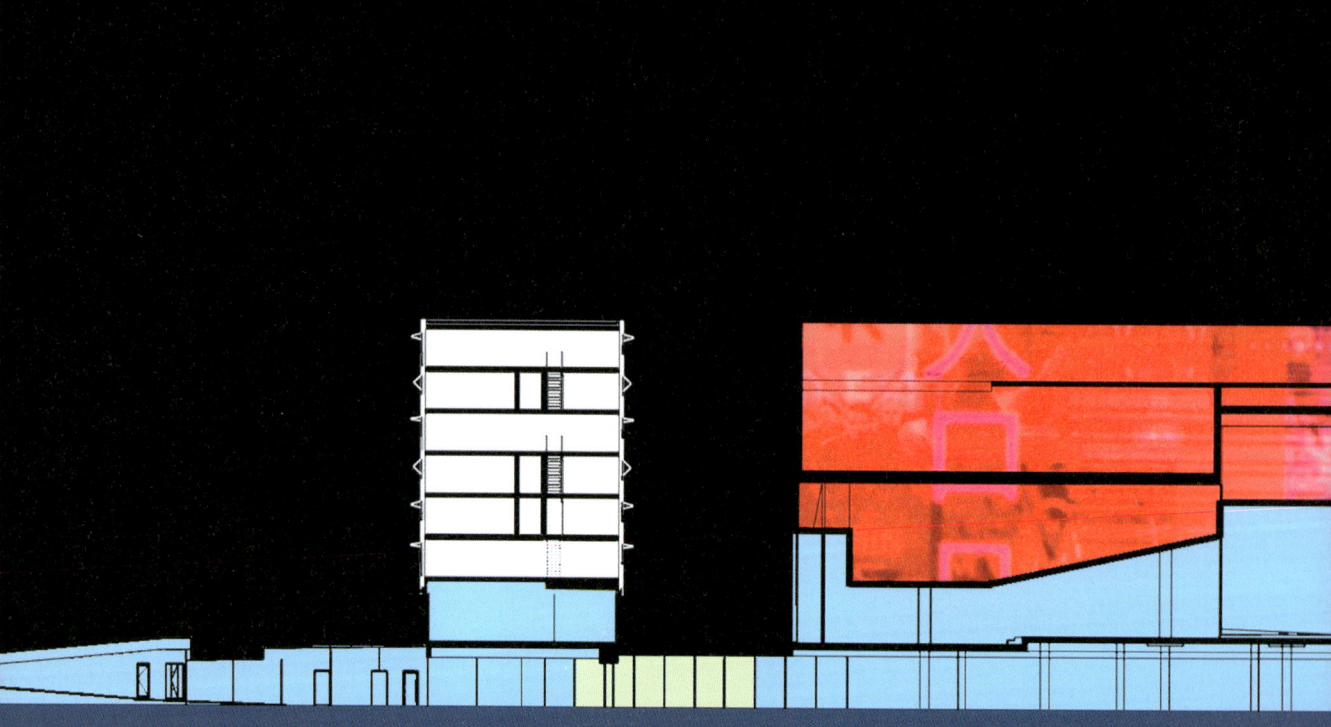

Tunisia
Position

New houses were designed for various locations in Tunisia for victims of the 1969 floods. In Abida the new houses are designed to match the traditional clay houses in scale and size. They consist of a main room and one or more side rooms. The position of each house was decided individually, in consultation with the families. Sometimes the houses were built on an open site at some distance from the other houses. However, new houses were also often built on the premises of houses that were already there, so old and new now stand amidst and alongside each other as equals.

Sporenburg
Position

In the plan by West 8 for the Borneo and Sporen-burg peninsulas in Amsterdam's Eastern Harbour District, various buildings were developed for the ends of the blocks. These terminations of the linear series of dwellings adopt a different orientation to these series. They are smoothly inserted into the three-storey volume, but do not coordinate with the system. They introduce a different typology – the house with its own front step – and thus follow their own logic, suggested by the situation and the oblique lines. The terminations are the only disturbance in the homogeneous neutrality of Sporenburg. All the end buildings belong together, though they differ in size and angularity. Because of the oblique line, the end blocks break free from the extended rows, creating an intermediate space – also collective – that gives the dwellings a second orientation.

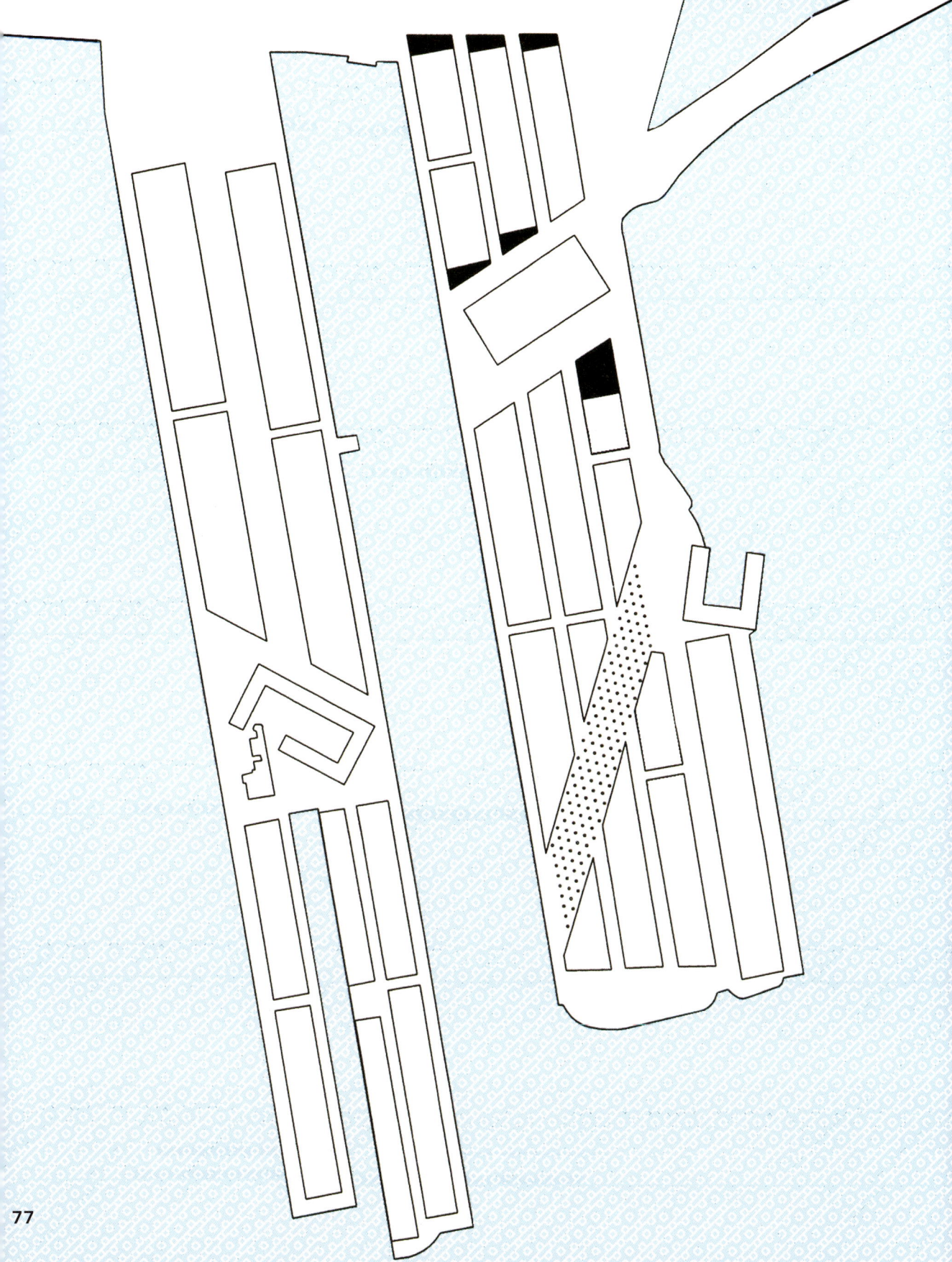

Ichthus
Position

This building in Rotterdam's city centre has a multi-faceted relationship with the city and the urban activities surrounding it. It is an autonomous object with a complex programme that creates an exceptional spot in the city but is simultaneously interwoven with it, owing to the omni—directionality and the urban functions at street level. • Above this there is a wide range of housing types, with various kinds of apartment, above which there are generously glazed penthouses amidst a field of flowers. As a figure the building has an egalitarian understanding with the neighbouring Laurenskerk. This church is an important reference, but not the only element of the context in which the Icthus project assumes its place. The height of the surrounding construction played a decisive role as well.

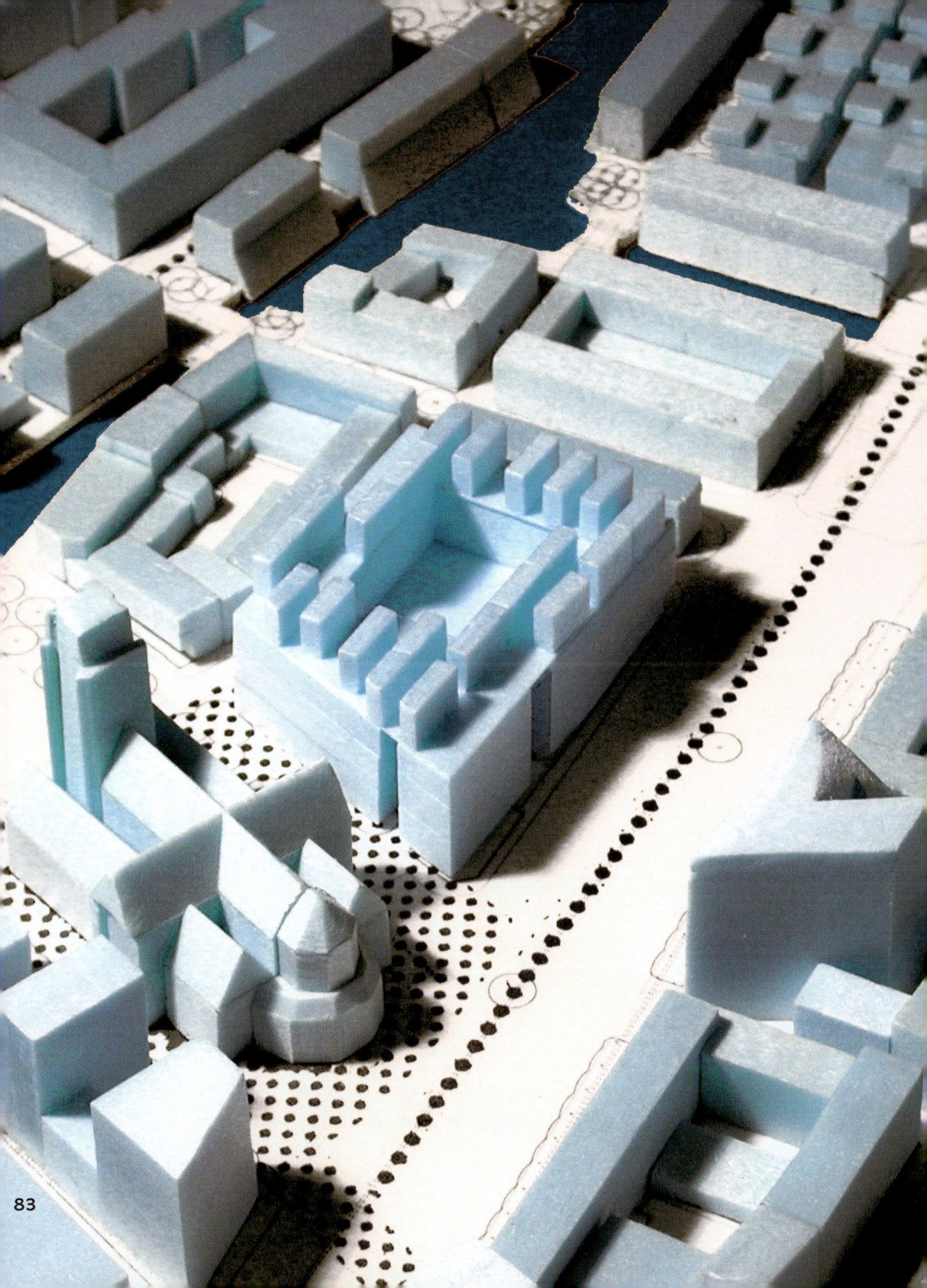

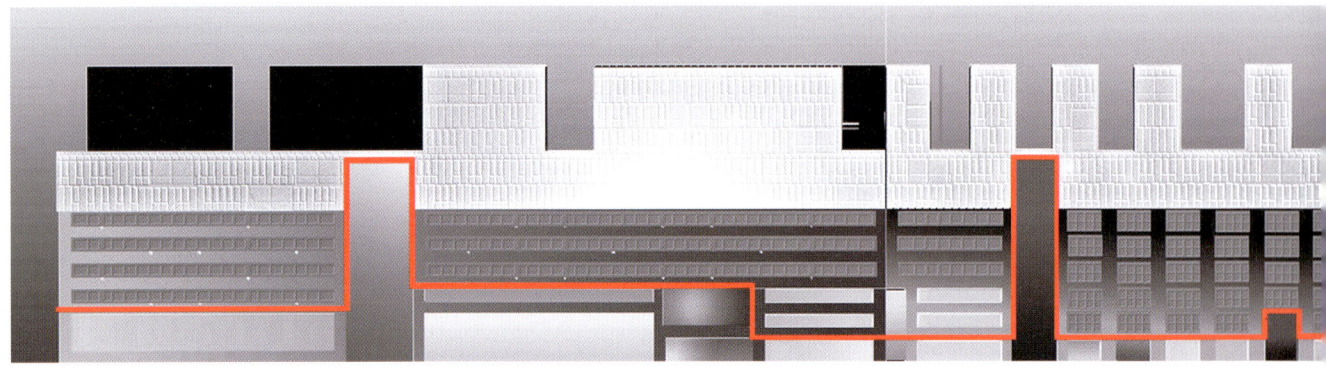

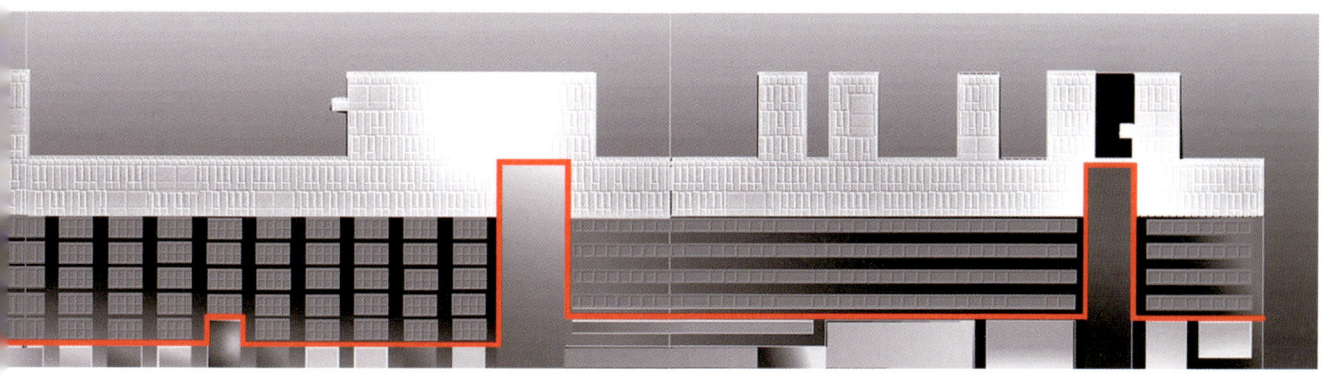

Block 30
Position

These eight compact buildings, each with five dwellings, stand like 'book ends' in Block 30, part of Amsterdam's new district of IJburg. The book ends stand on each side of volumes that are inserted into the block: a post office building, a school and two residential blocks. The book ends are identical but do not form a symmetrical pair.

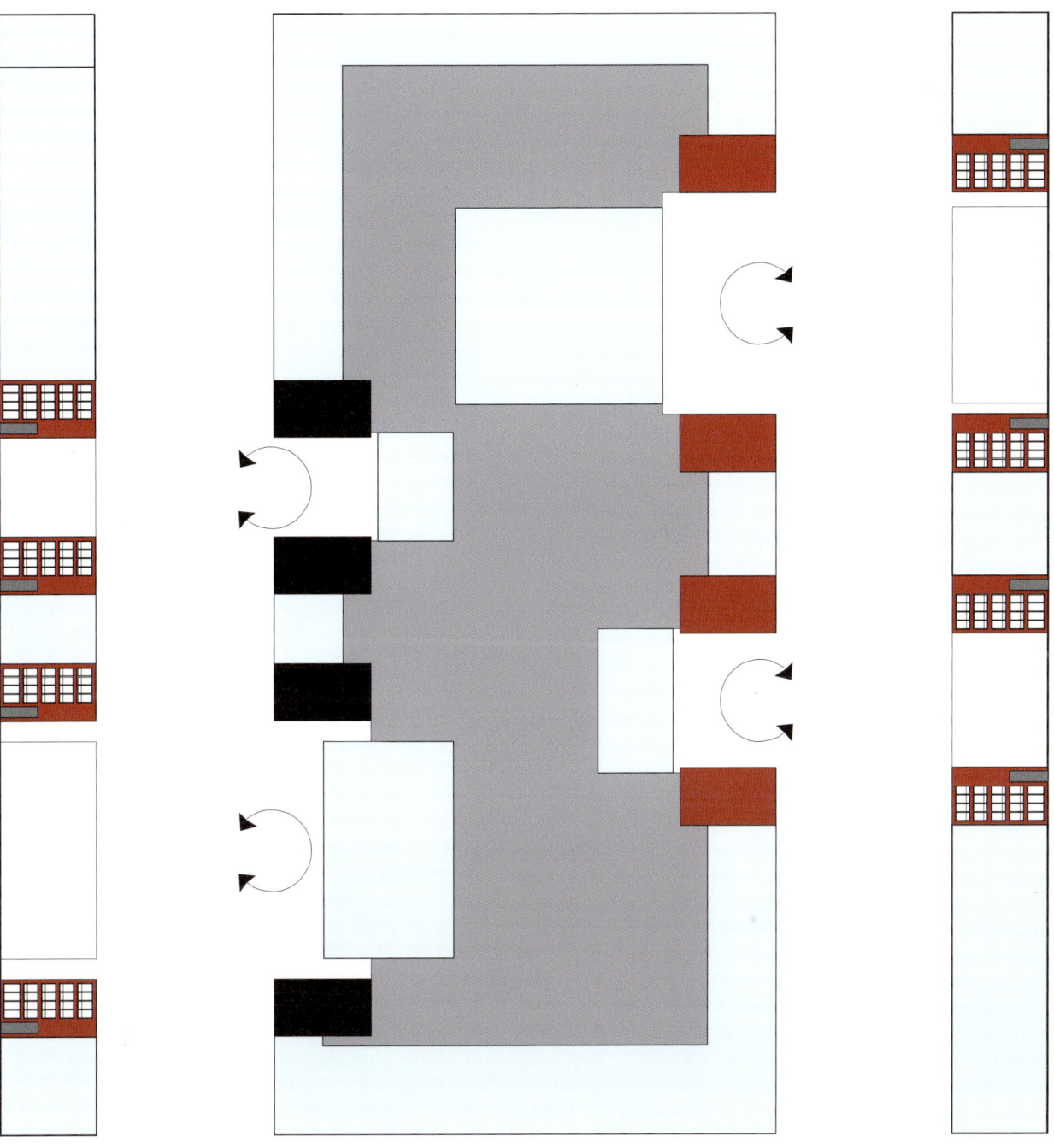

Suytkade
Position

A residential building on the water was designed for a new district in Helmond to an urban plan by Soeters Van Eldonk Ponec. It forms the entrance to the two residential buildings behind and also contains the entrance to the communal car park. The building is a single volume with an omni-directional orientation into which 'holes' have been hewn. The floor plans are organized in the same way every two storeys, but their orientation also changes every two storeys, which contributes to the 'omni-directional' character of the figure.

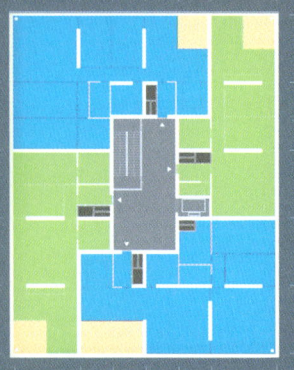

6e verdieping

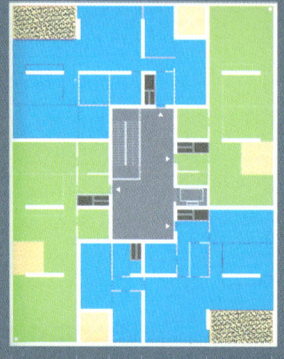

7e verdieping

8e verdieping

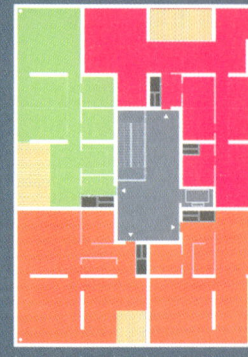

9e verdieping

2e verdieping

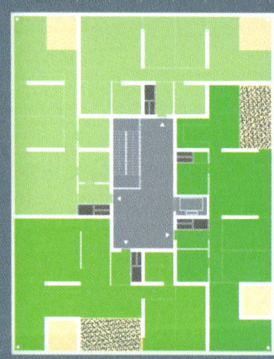

3e verdieping

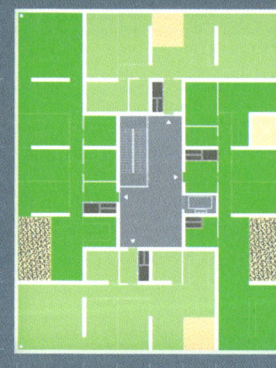

4e verdieping

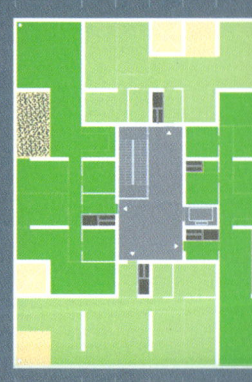

5e verdieping

parkeerlaag

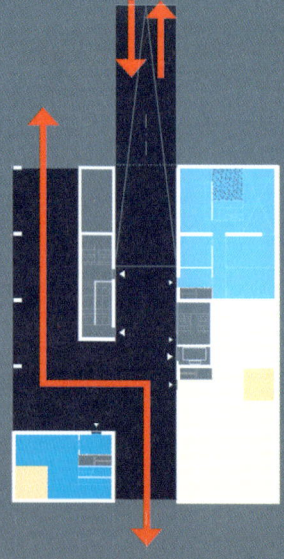

begane grond

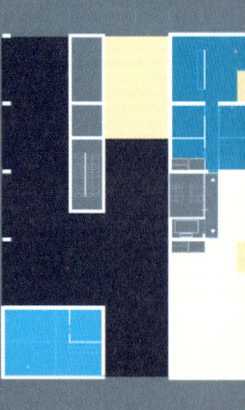

1e verdieping

Oosterparkstraat
Position

The houses in the Tweede Oosterparkstraat in Amsterdam form a street frontage, an extended building in a long and narrow nineteenth–century street. In a setting that has been fragmented by urban regeneration, the project reinstates a nine-teenth–century urban order. Rather than reading it as a repetitive series of houses, the project was treated as a building with a central entrance, por-ches and galleries that concludes in a true termin-ation to each flank. However, it also has a unit-wise articulation, with two stacked maisonettes and an apartment with a large private terrace above. • Viewing it from a range of perspectives it is therefore a component of the street elevation, an independent building and a succession of properties.

Nonnenveld
Position

The residential complex stands on the edge of the Chassé site in Breda, for which OMA devised the urban plan. The building occupies a position between the old and the modern city. It finishes an incomplete urban block, the old and new sections of this block forming a unit that can measure up to the individual objects in the Chassé Park. These independent buildings all differ in volume, size, height, colour and materials.

Erasmusgaarde
Position

In 'Field 17' of Zuidwest, a residential district in The Hague that is undergoing extensive redevelopment, the concept of position is significant on various levels. The residential buildings stand in relation to the ground level in a specific way. The Erasmusgaarde is not simply a field of grass with buildings on it. The buildings are part of the ground-level configuration and are related to it. The elevations extend in all four directions in the paving around the buildings. • The residential tower occupies an exceptional position within Field 17. The tower belongs to a different order and is of a different scale. The call to build a tower did not arise from the project for Field 17, but was already decided in the urban plan. The tower introduces an element between Field 17 and the city, an element that relates to the city, to the field as a whole and to the square at its foot.

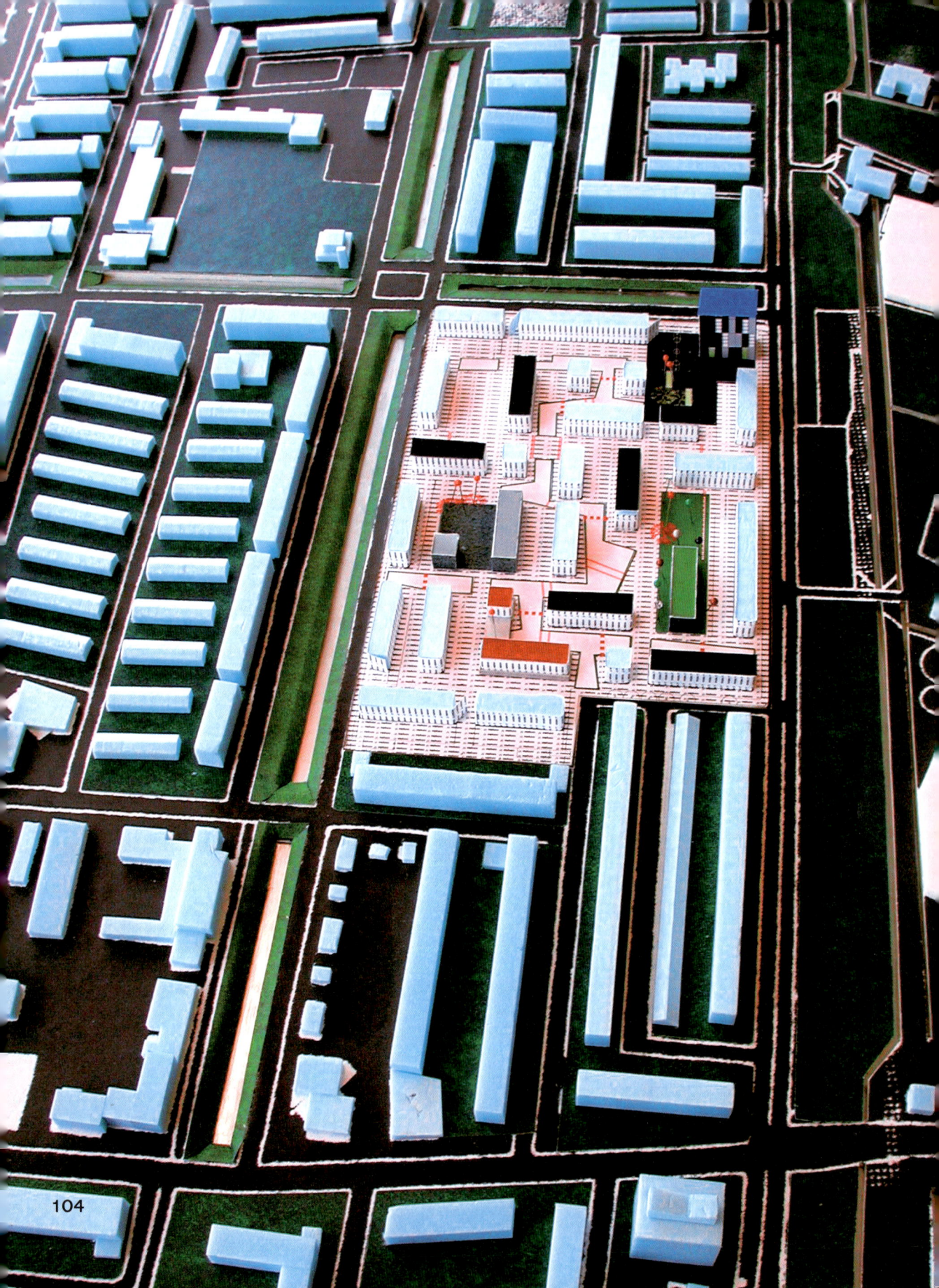

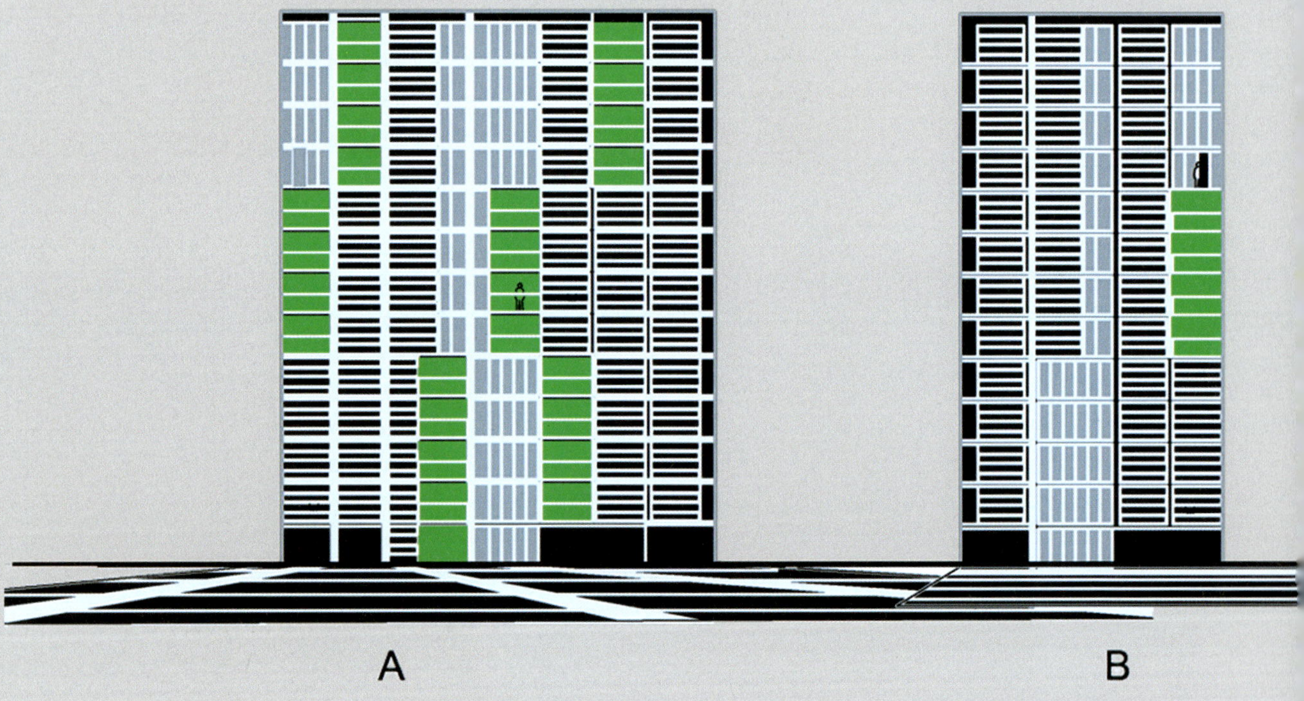

A

B

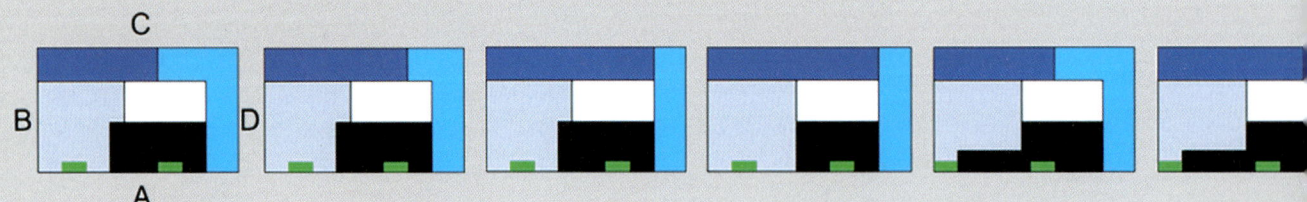

C

B

D

A

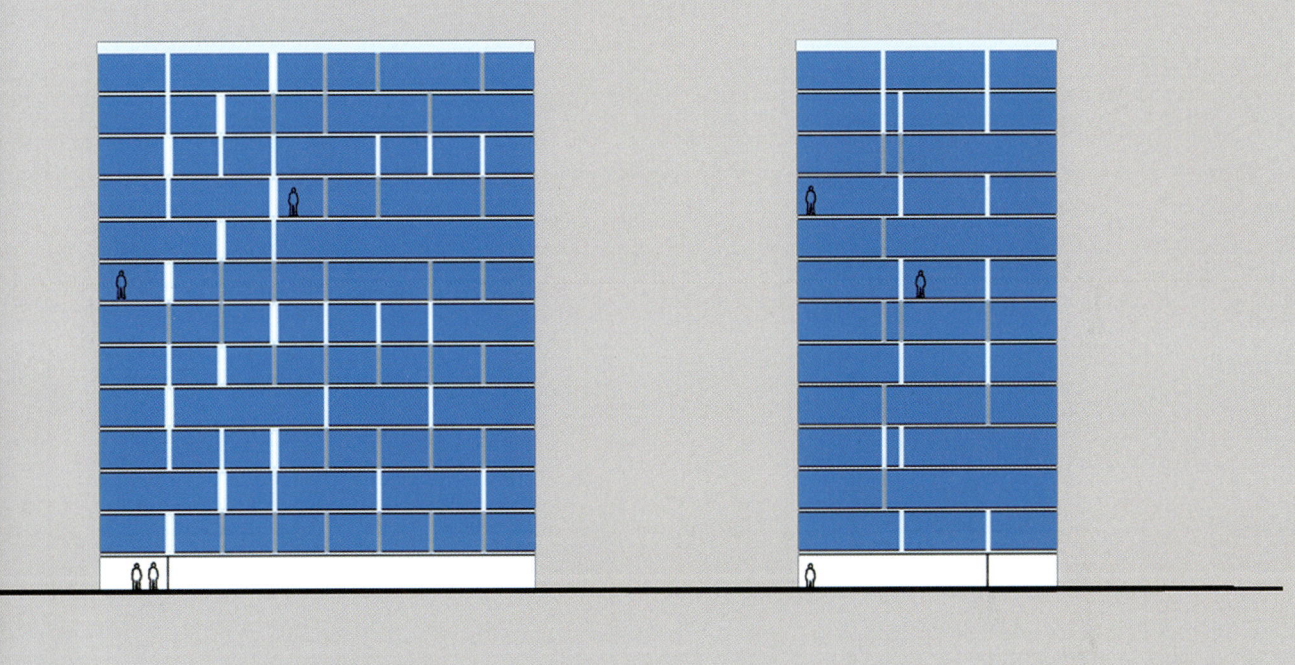

C   D

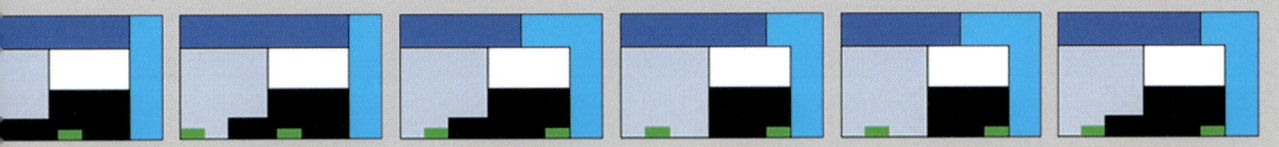

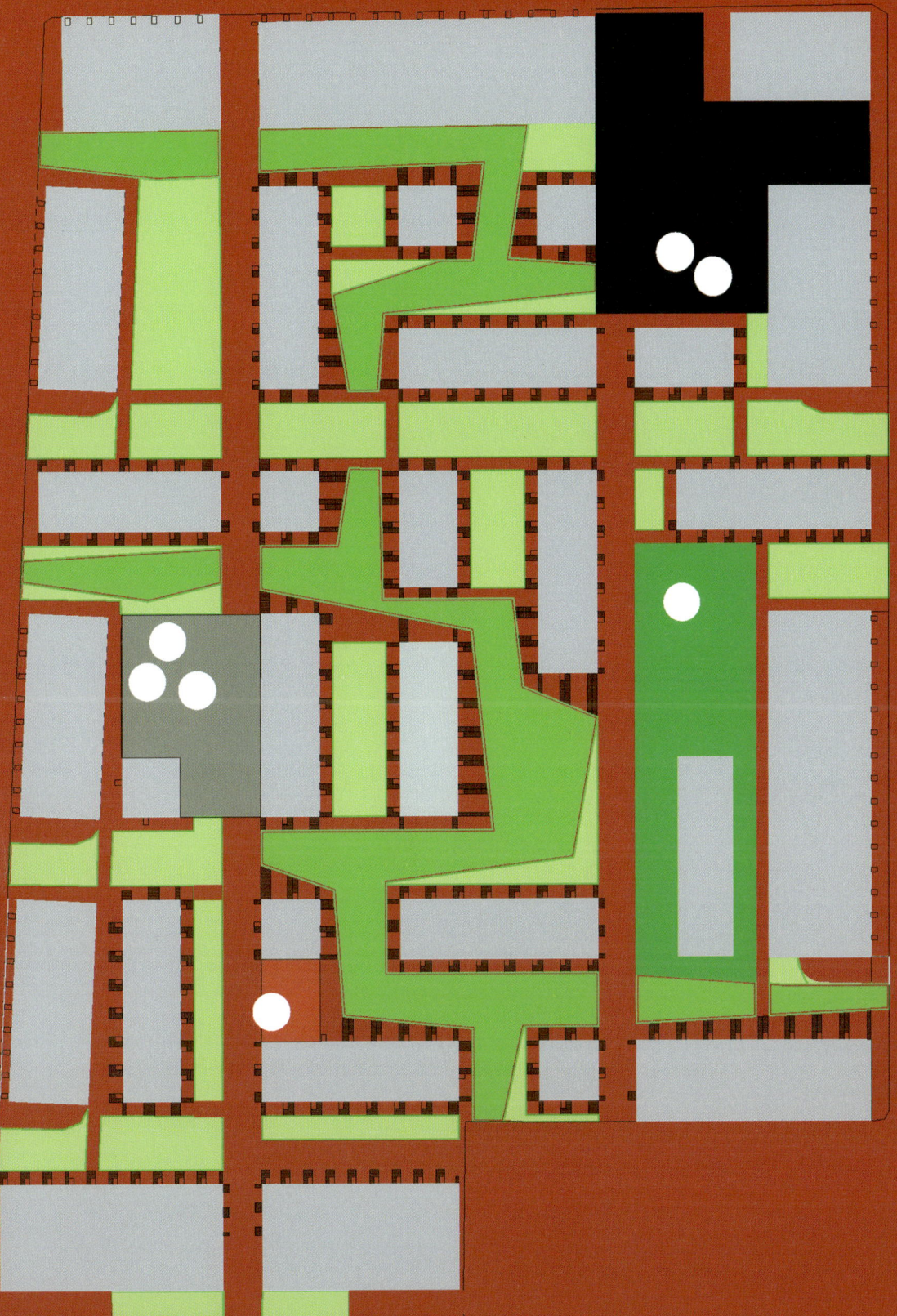

238
343

761

1040   3500

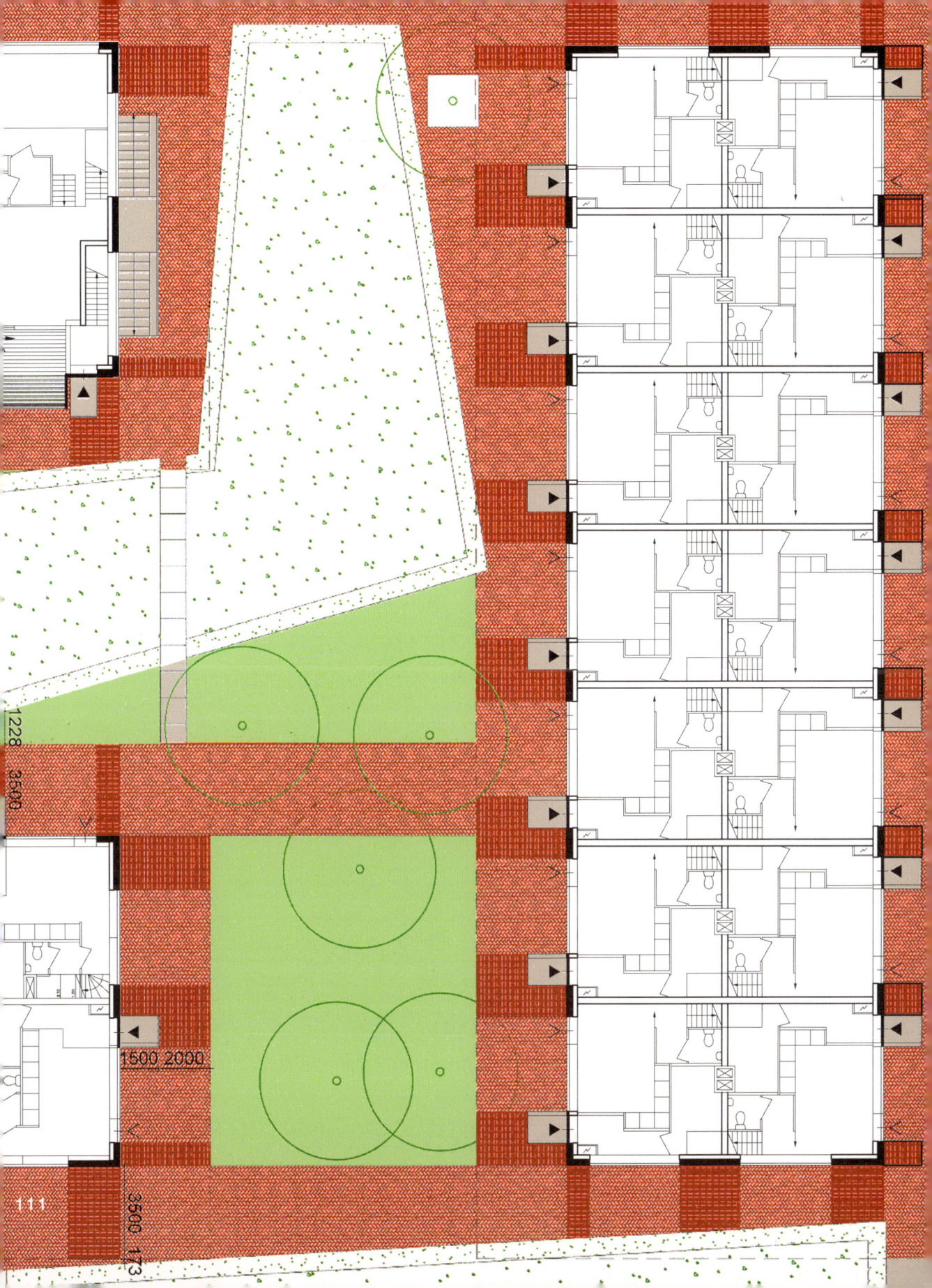

Muiderstraat
Position

In the office extension to the eighteenth—century monument on the corner of the Rapenburger-straat and Muiderstraat in Amsterdam, the new is not dependent on the old and nor is the monument reduced to the representative facade of the new building. Old and new are equals and stand beside each other more or less autonomously. Since the new construction literally wraps around the monument, old and new are conjoined to form a single object around a raised inner court-yard. This object is a match for the two substantial objects standing opposite: the synagogue and the Netherlands Film and Television Academy.

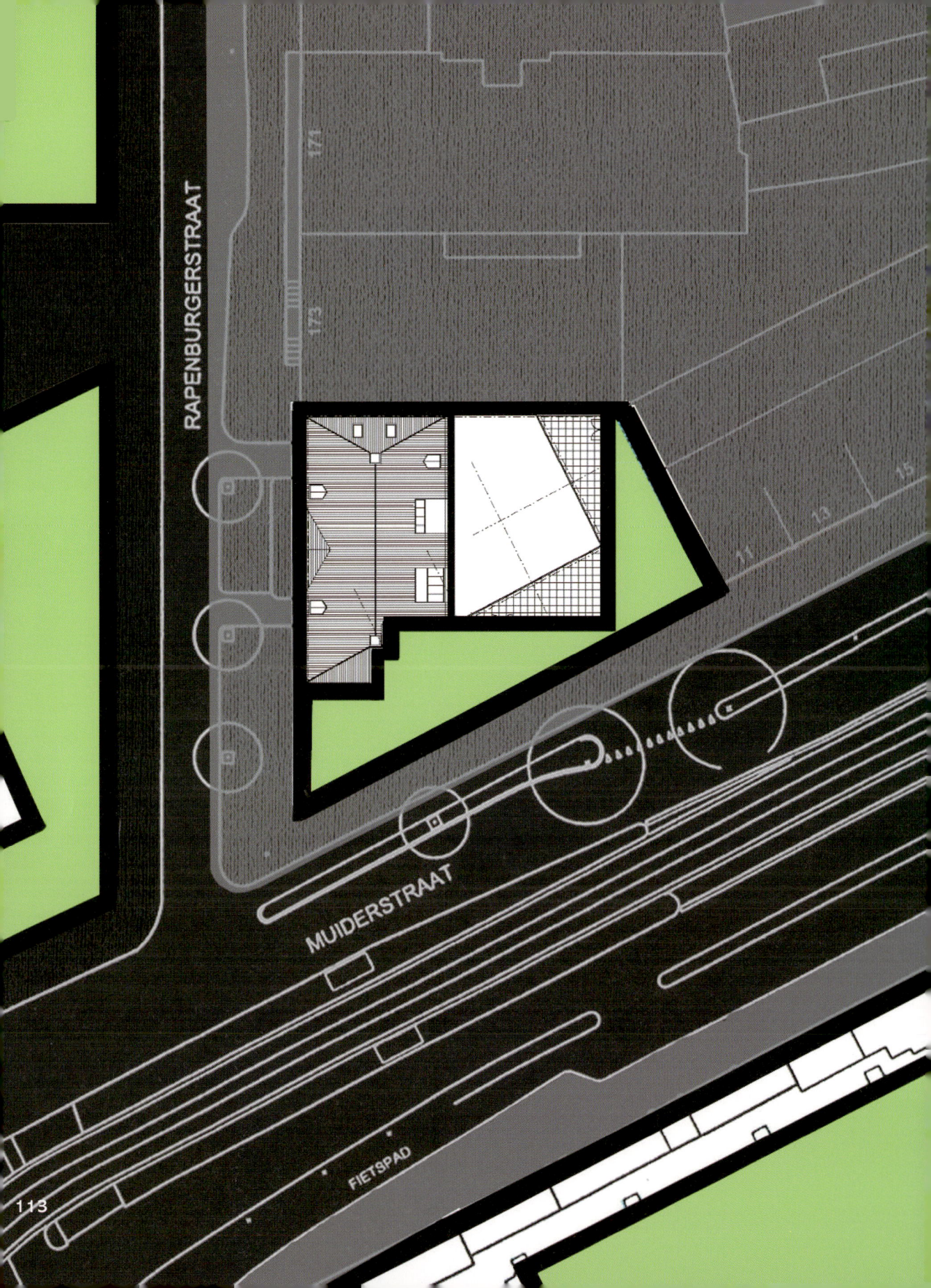

RAPENBURGERSTRAAT

MUIDERSTRAAT

FIETSPAD

113

Time

Time is a factor that features in various ways in the positioning of my architecture. Considering the temporal dimension in architecture and urban planning provides anchoring points to interconnect the pre-existing and the new. For me it is also about the question of what there is – or has been – that might be continued, as well as what is absent and must therefore be added. For me it is not only about revealing a relationship between present and past, but also about establishing a relationship with the future. What there was, what there is and what could be there – all three are of equal importance.

Time

Oosterpark
Time

The street frontage opposite the Oosterpark in Amsterdam was developed about a century ago, resulting in it being cohesive because all the facades of the single and double plots were constructed in accordance with a system of implicit rules, the product of erstwhile conventions and habits. The project links in with this by specifying and applying these 100–year–old rules — and the exceptions to them. That gives this project a self-evident clarity that is capable of bridging a century in time. • The typology also corresponds with the street elevation, consisting of two stacked duplex apartments. This typology is recapitulated in the project's ground–floor dwellings. For the topmost units, double–width loft–like apartments were introduced, which are unusual for this neighbourhood. A fifth storey was added in order to make the project economically viable.

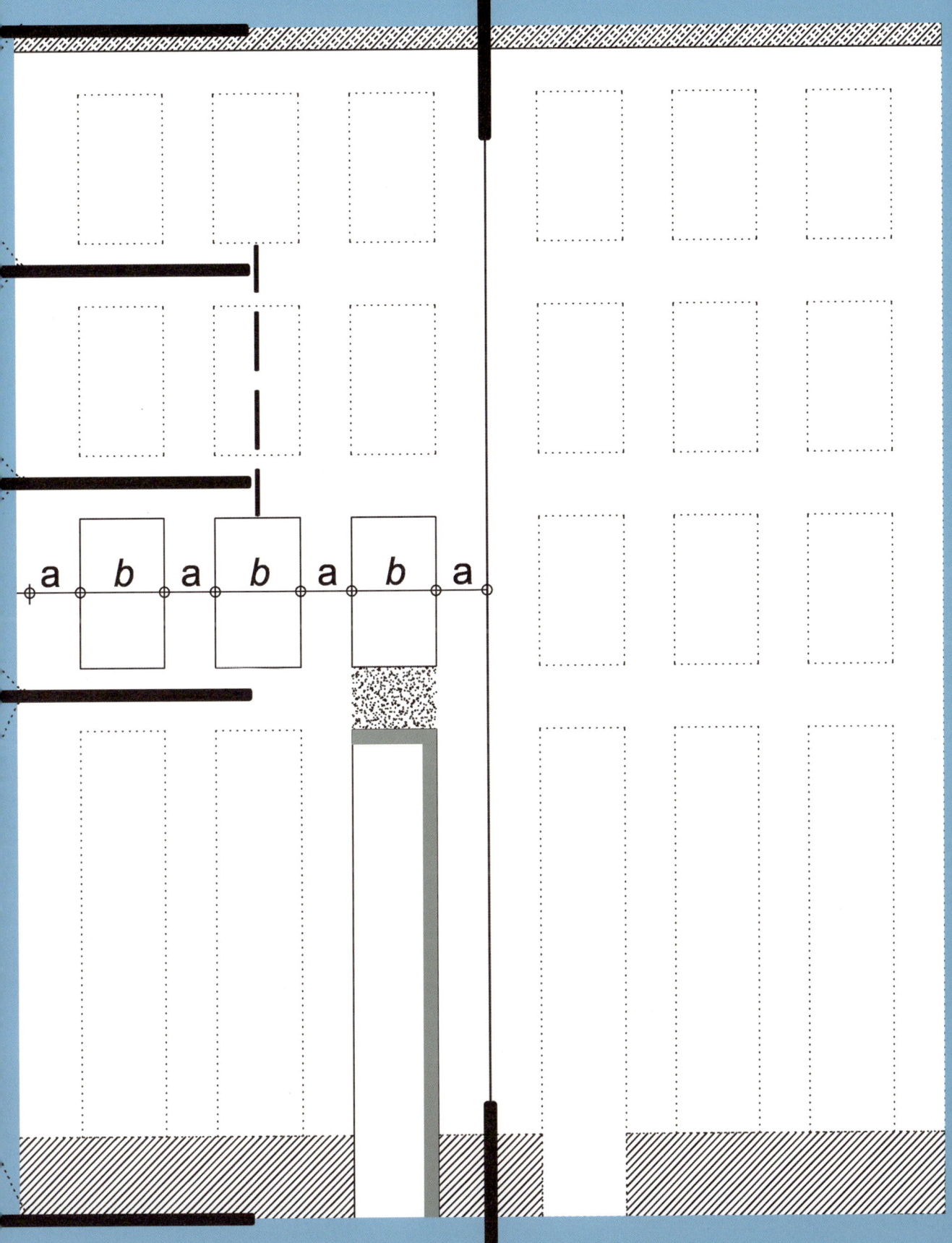

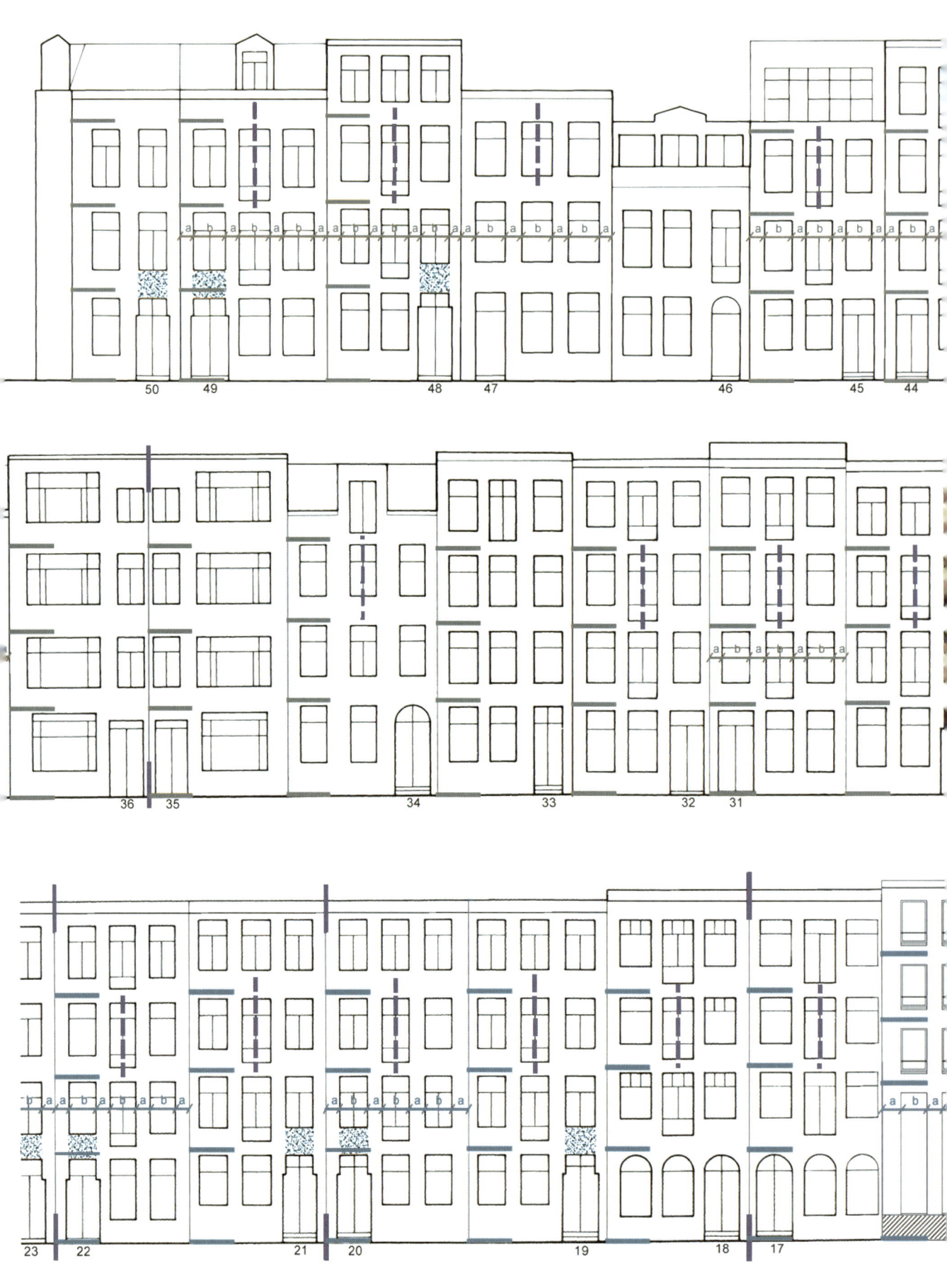

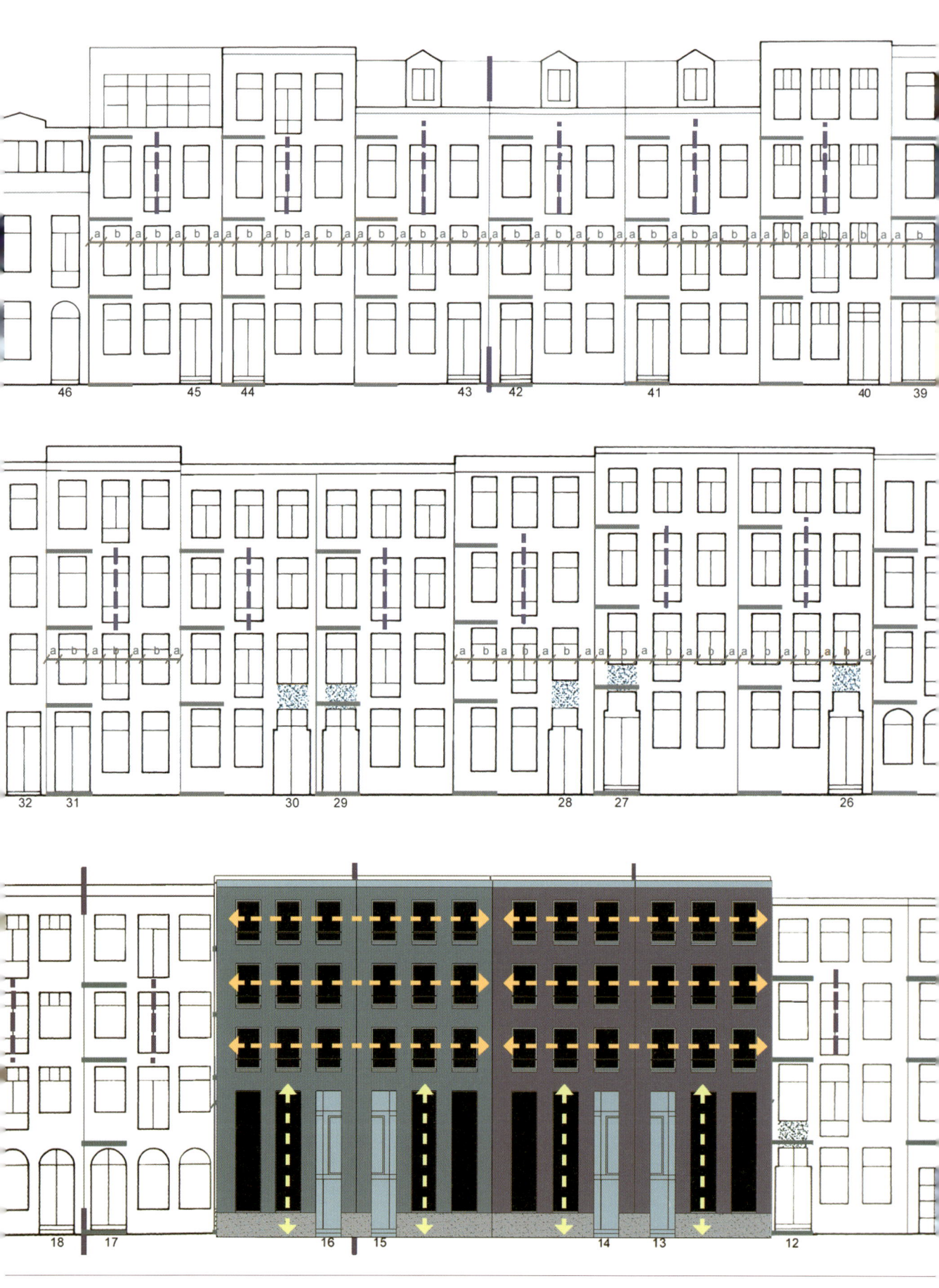

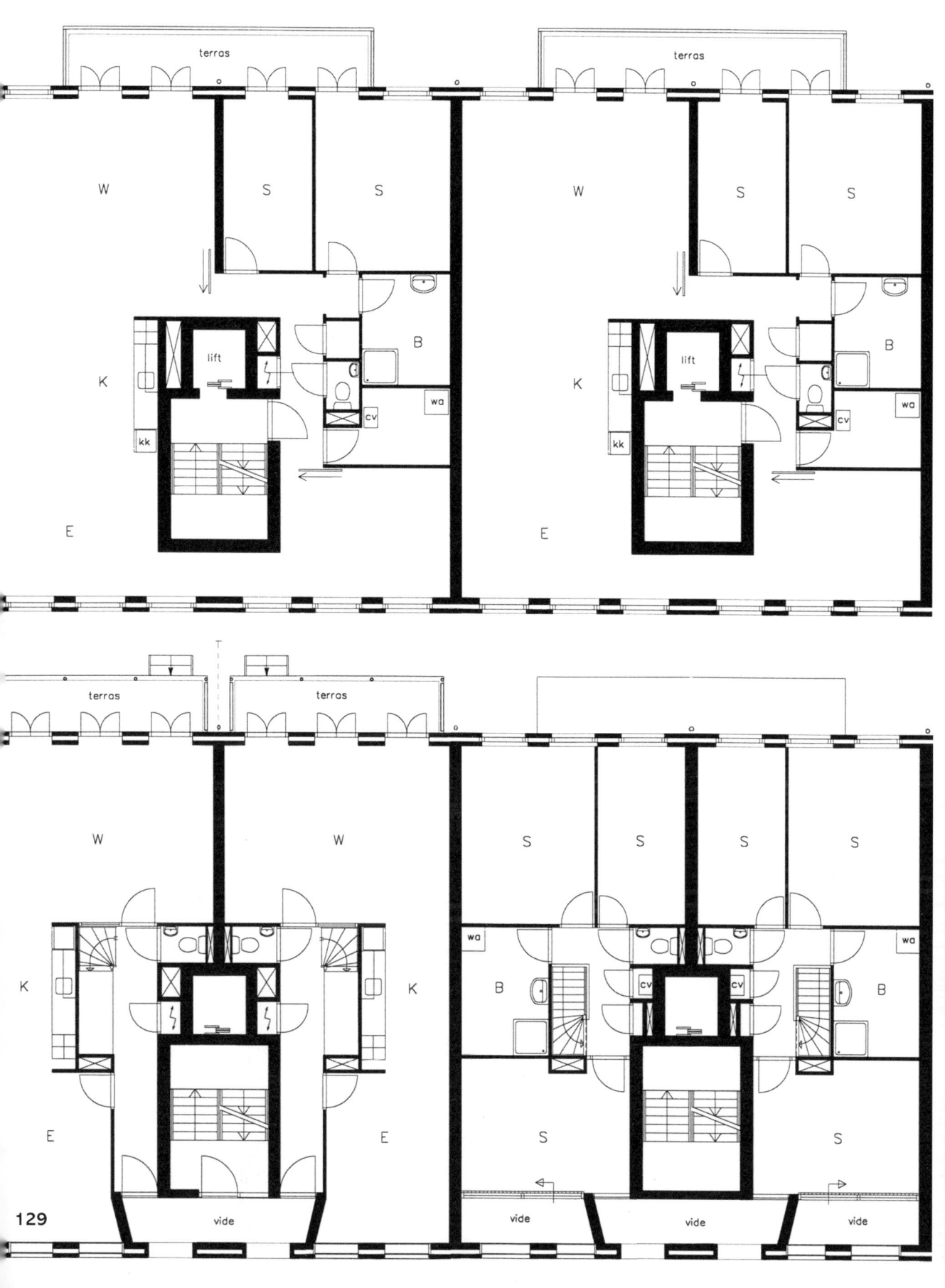

Pendrecht
Time

For the transformation of a zone on the south side of the Rotterdam district of Pendrecht, Lotte Stam-Beese's original plan from the 1950s proffered the solution. Her balance of construction and greenery is retained, while the obscured distinction between main road and child–friendly street is reinstated. The parking lots in Stam-Beese's plan are now an entrance to the half–submerged garages. This reinforces the typological mix within the neighbourhoods, creating a 'living unit', to use the terminology of that time. Part of the construction is retained, while the rest has been replaced with new blocks that are taller but remain within the street alignments of Stam-Beese's original plan and are therefore integrated into the typical orthogonal order of Pendrecht.

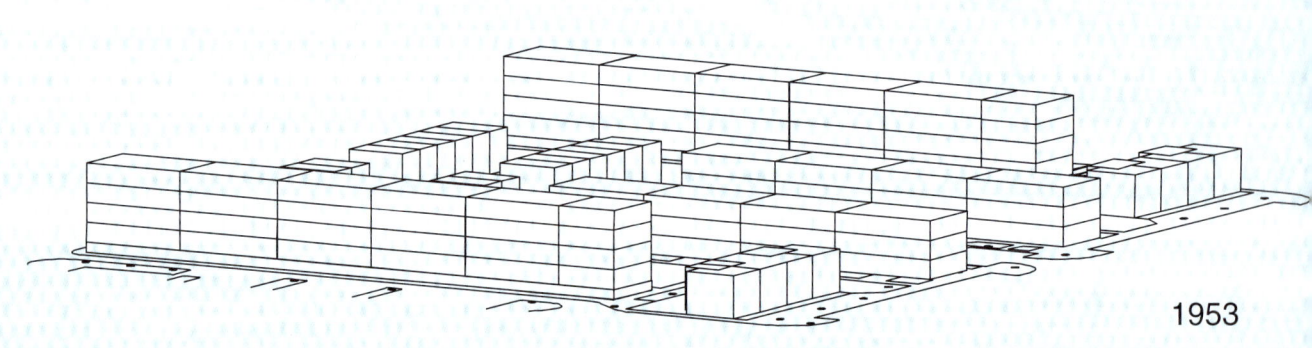

1953

2004

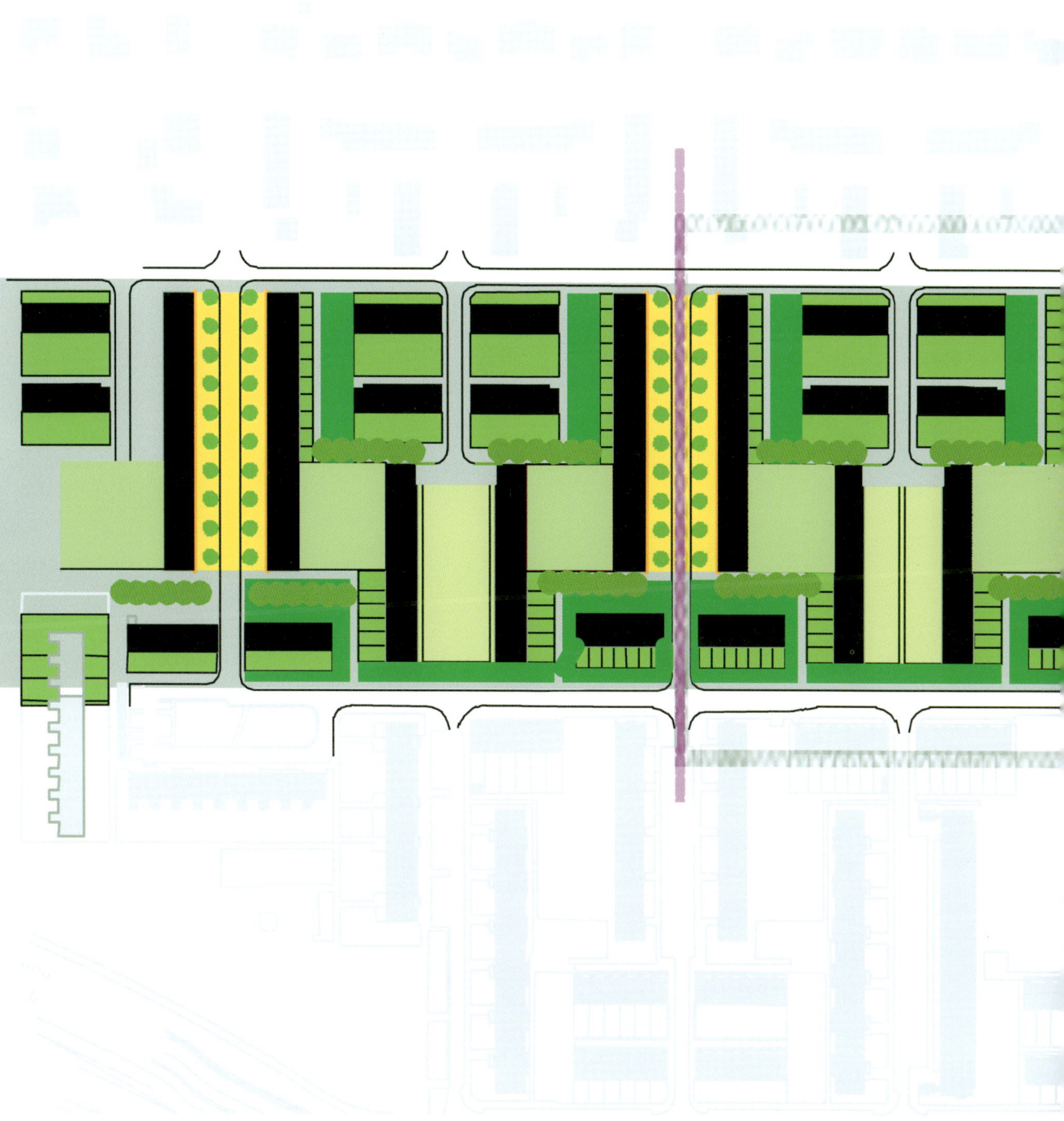

133

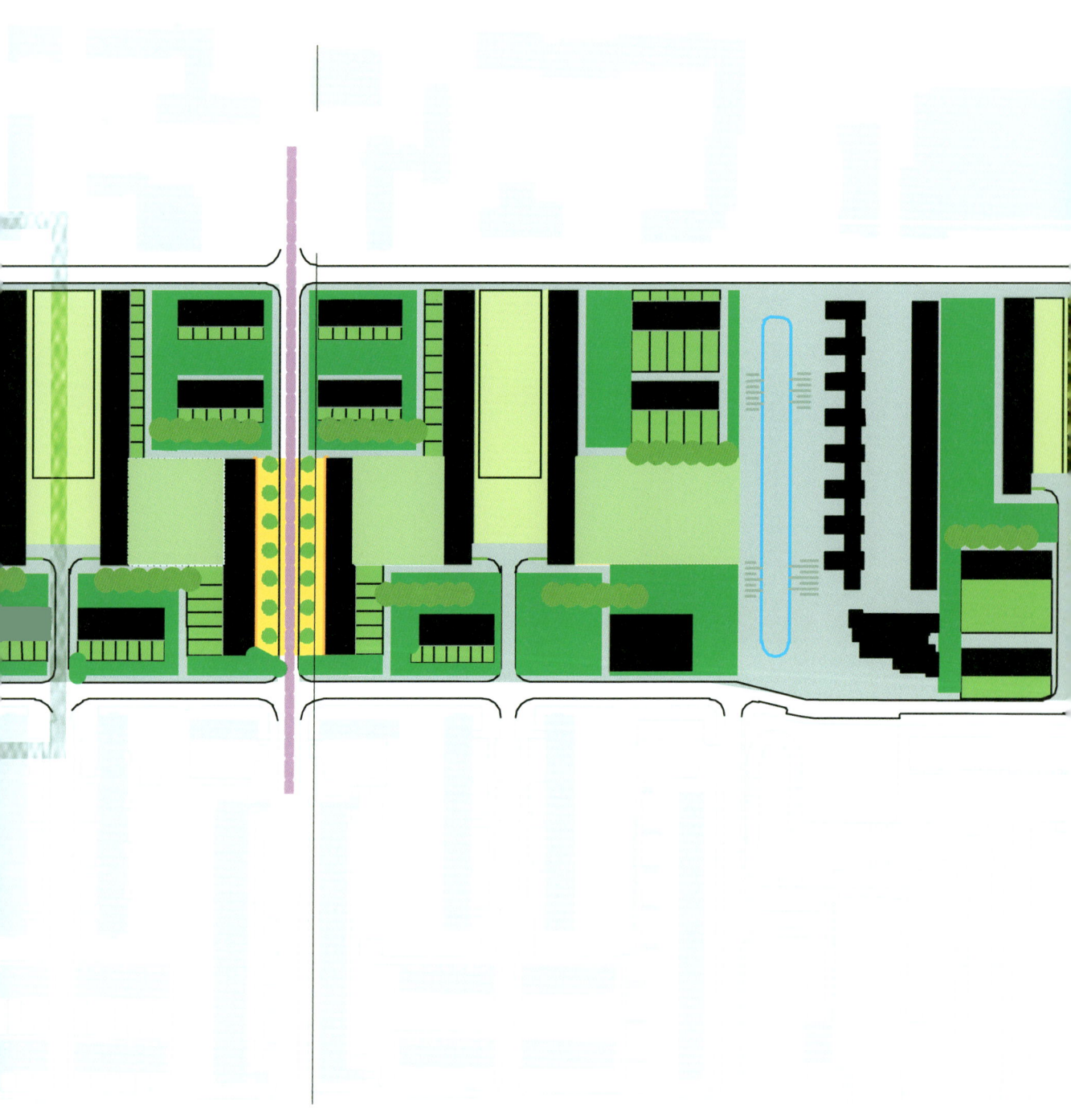

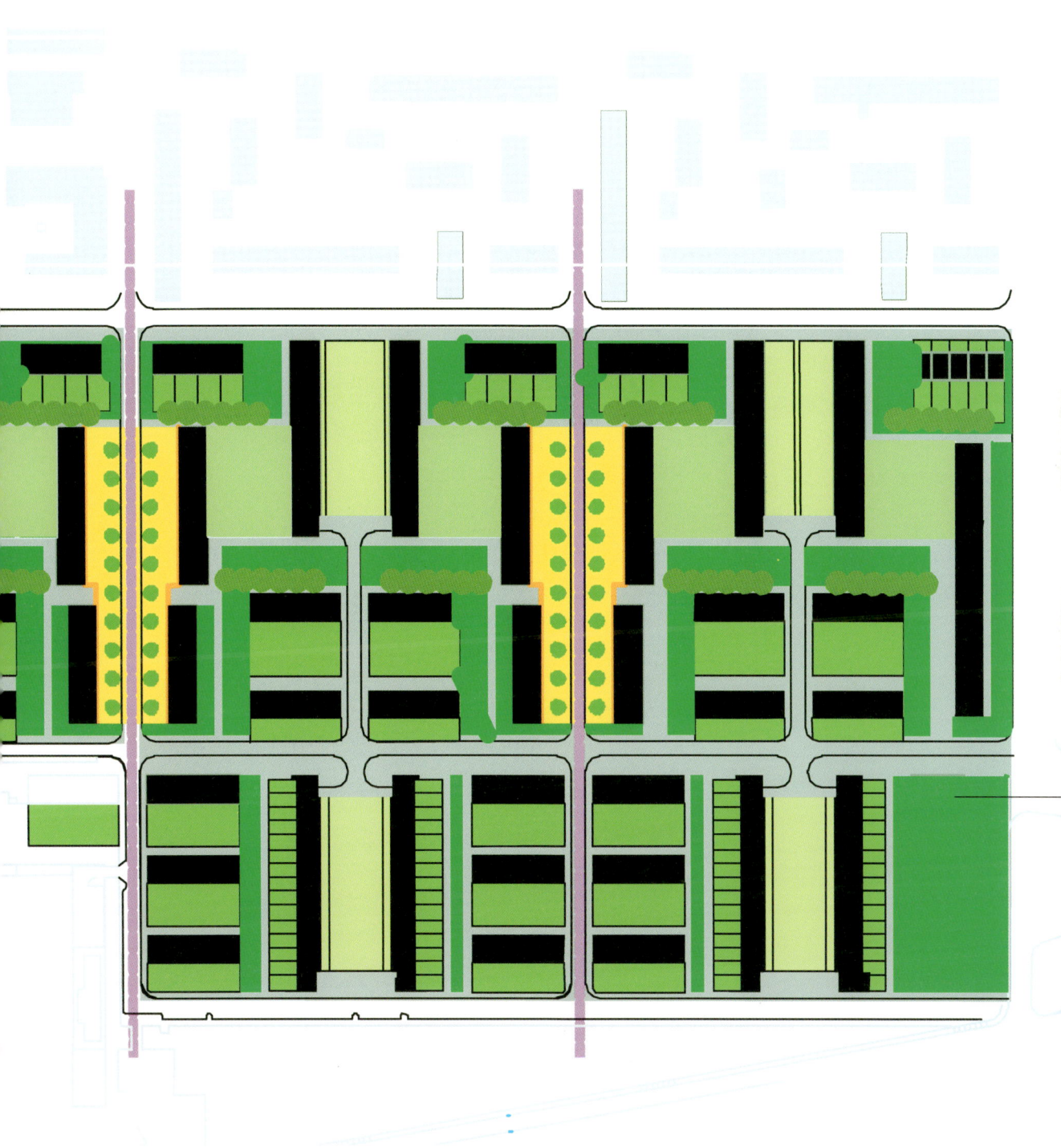

Muiderstraat
Time

Like the eighteenth–century house on the Rapenburgerstraat in Amsterdam, the new building on the corner of the Muiderstraat represents a moment in history, but without any allusion to the past. The facades of old and new buildings stand shoulder to shoulder yet seemingly indifferent to one another, without any formal or compositional resemblance. There is still a link between old and new, since the new facade, like the old, is highly transparent. Its ceramic louvres effect a filtering of light similar to that of the windows of the historic building, which are subdivided into little panes.

Ichthus
Time

Building alongside the Laurenskerk church in Rotterdam, amid surroundings that were reconstructed after the Second World War, inevitably broaches the factor of time. The pre–war city, the post-war city and the city of the future do not merely appear in various morphological layers, but also in the succession of construction heights (12, 18, 25, 35 and 60 metres). The main structure of the building is based on the succession of morphological layers and the height of the buildings.

1850   1930

MORFOLOGIE

RUIMTE

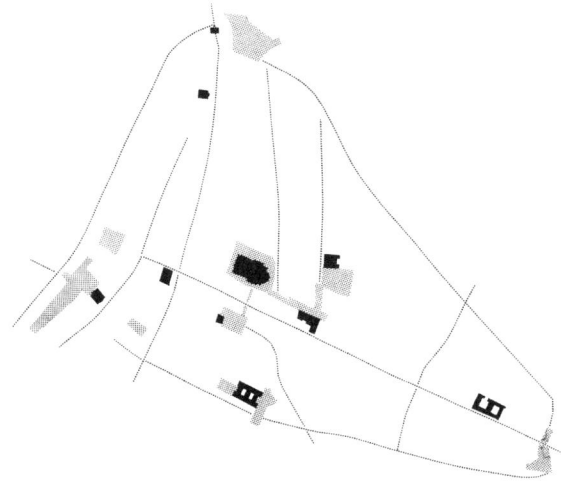
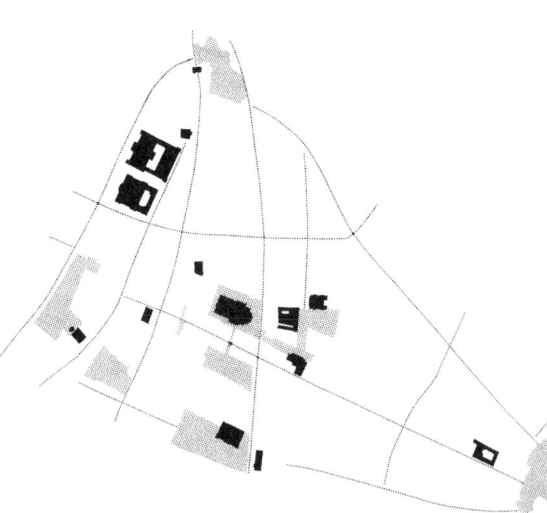

STRUCTUUR

1945　　　　　　　　　　1995

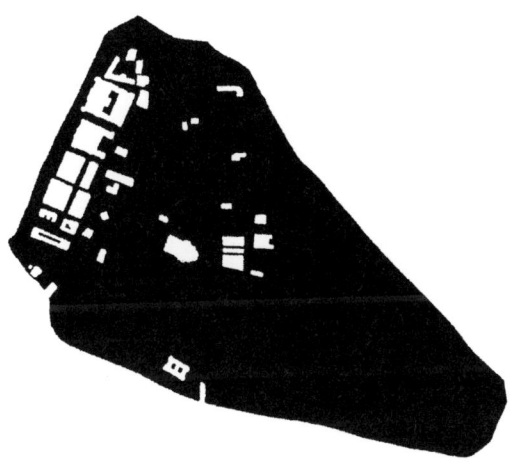 

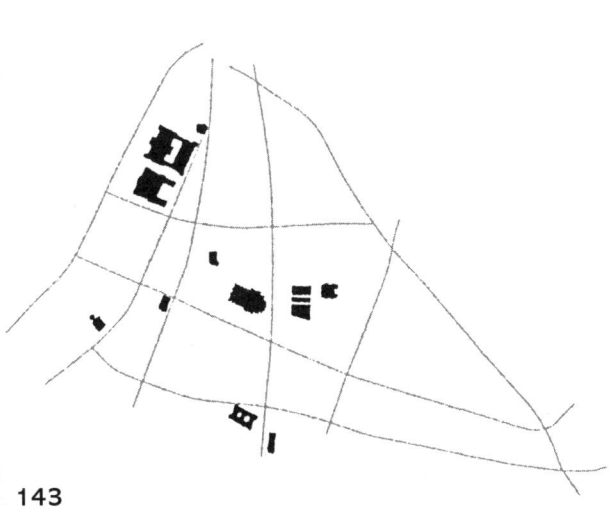 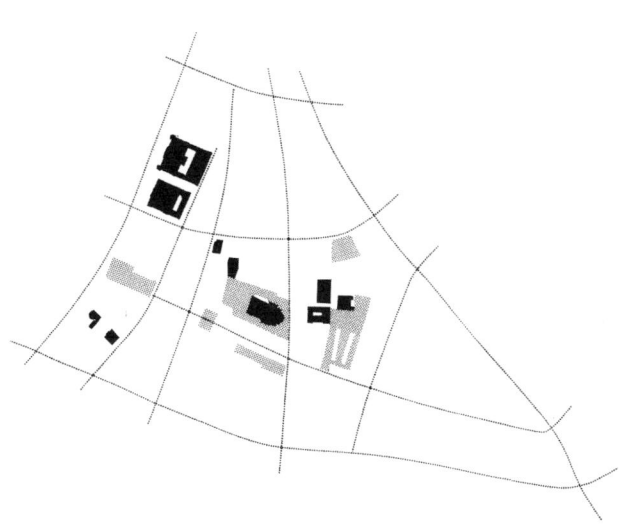

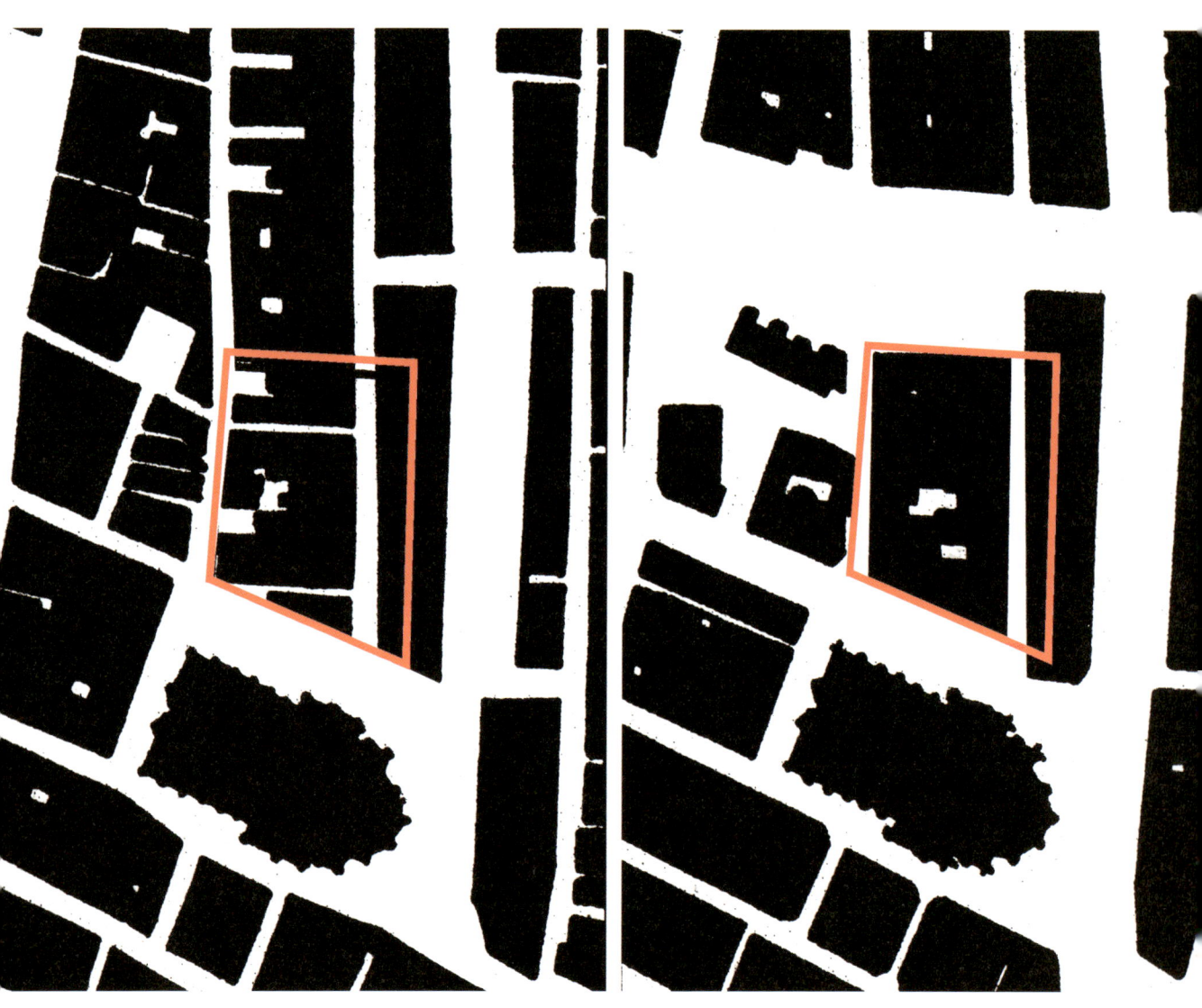

1850   1930

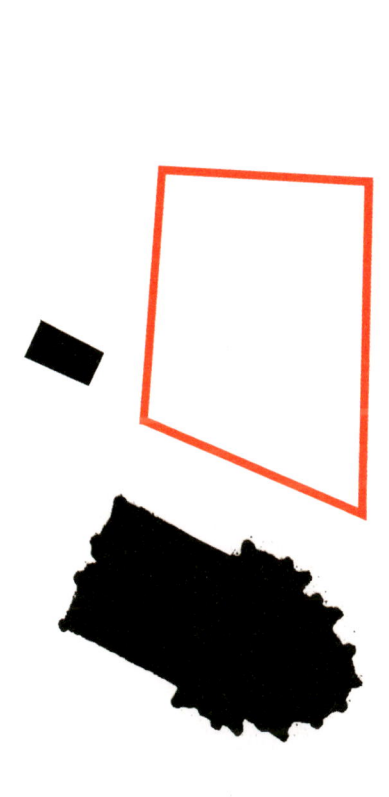

1945

1995

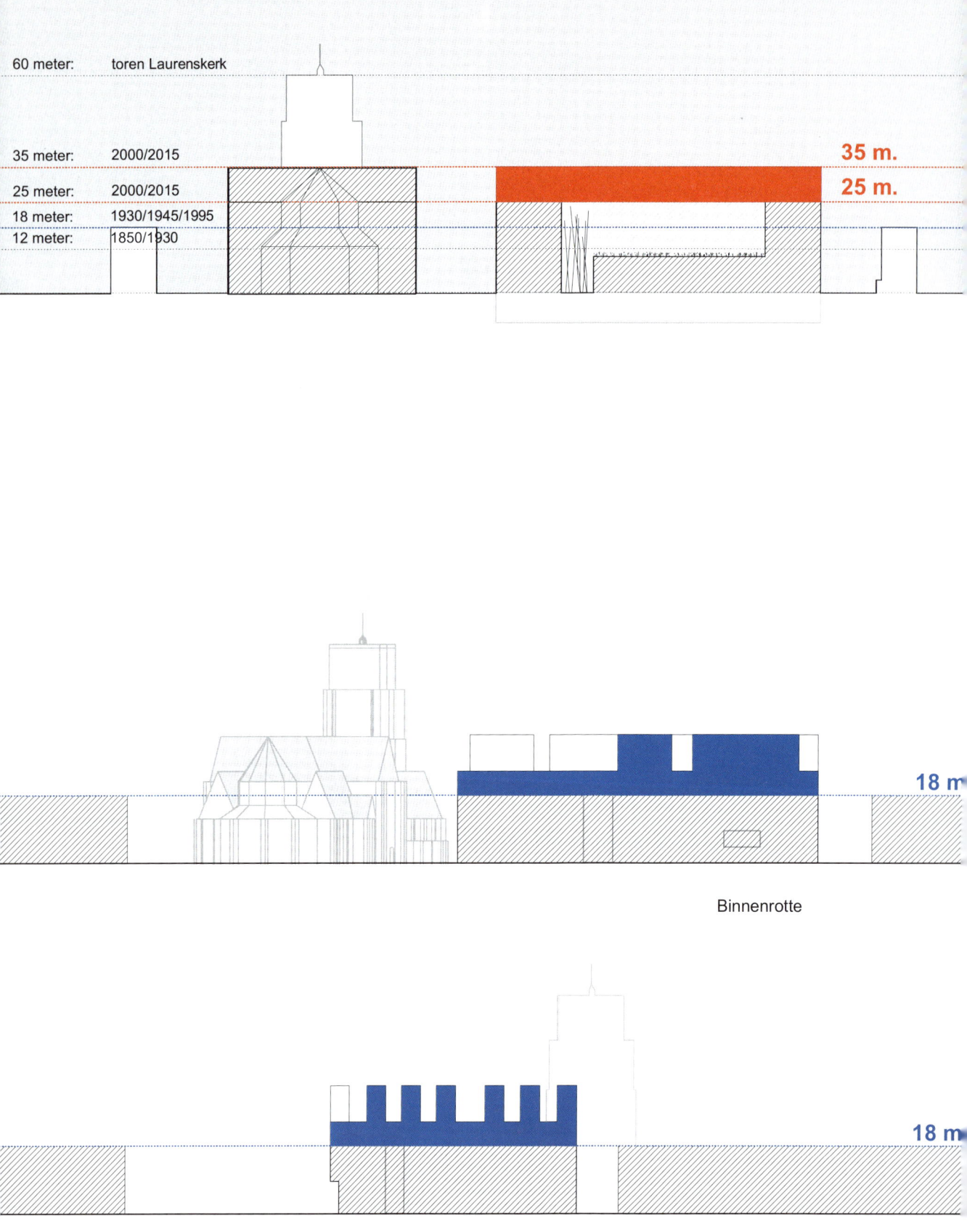

60 meter: toren Laurenskerk

35 meter: 2000/2015   **35 m.**
25 meter: 2000/2015   **25 m.**
18 meter: 1930/1945/1995
12 meter: 1850/1930

Binnenrotte

18 m

Bagijnenstraat

18 m

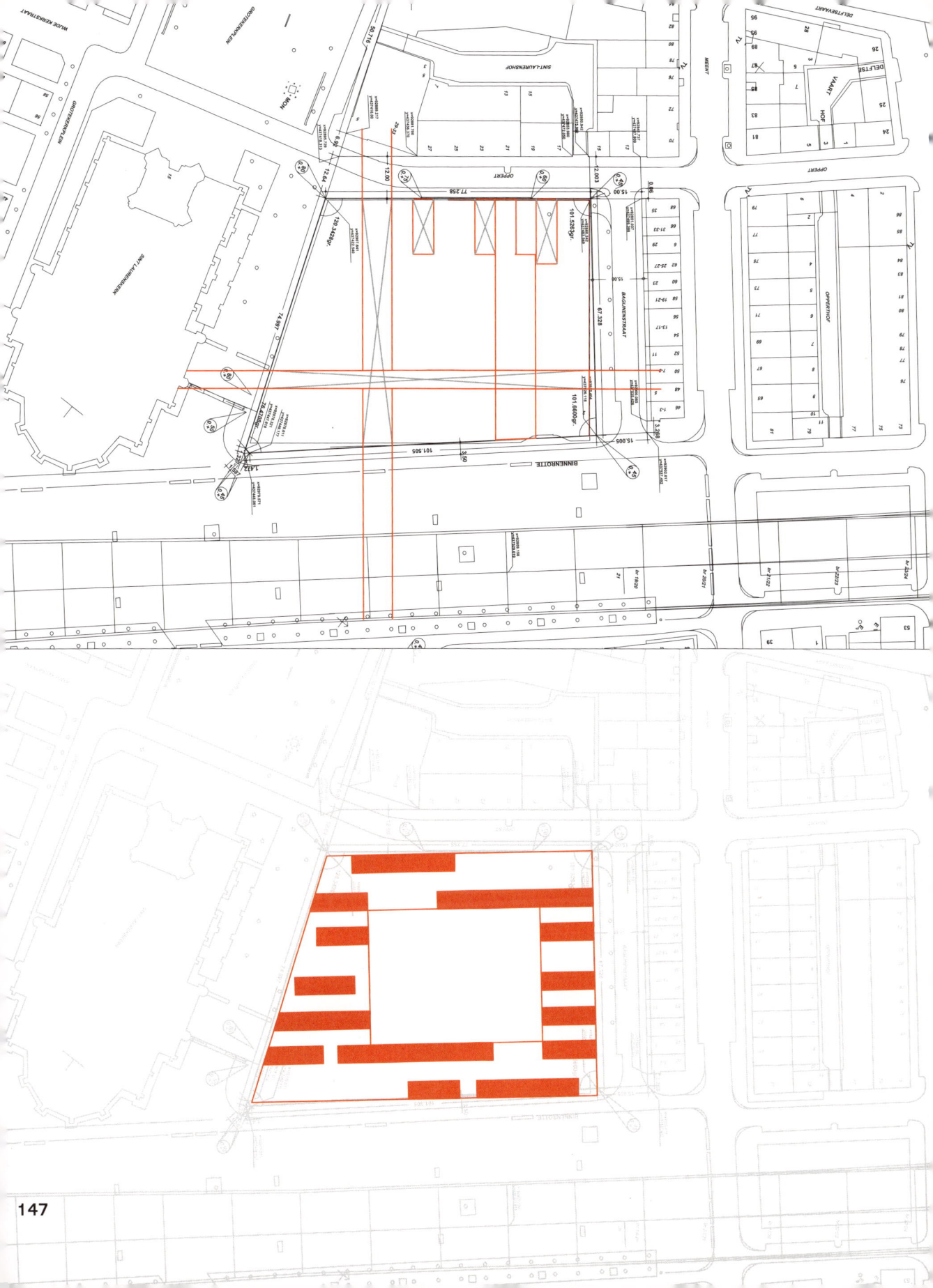

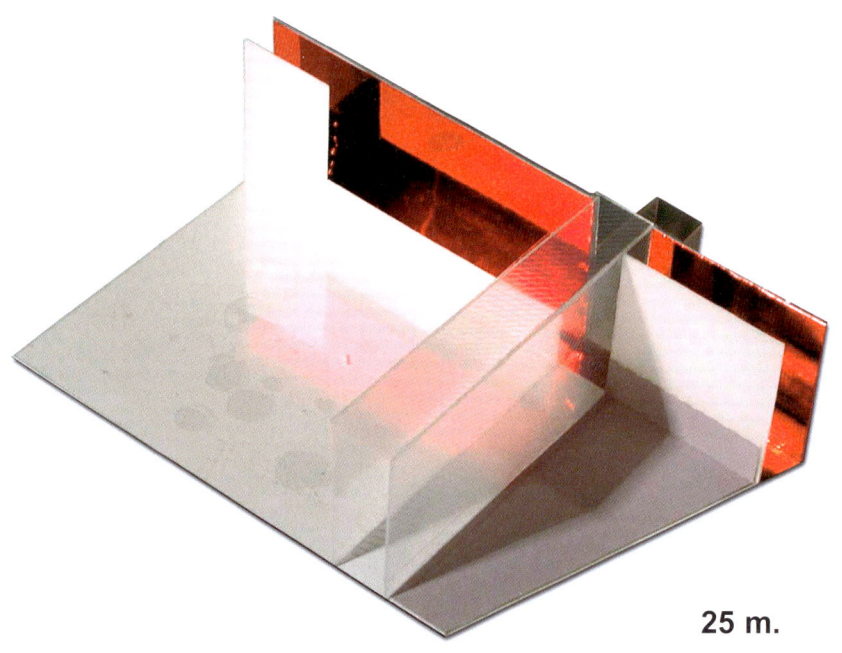

25 m.

25 m.

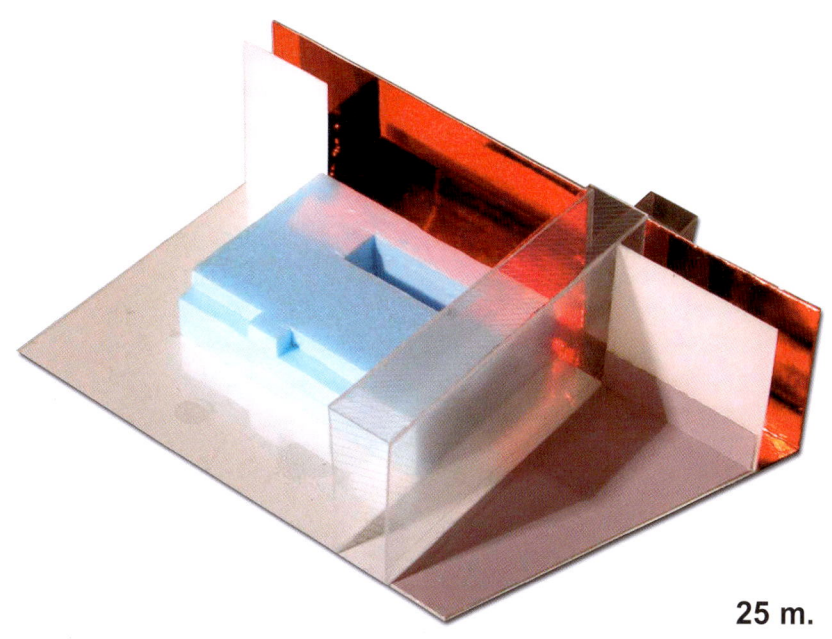

25 m.

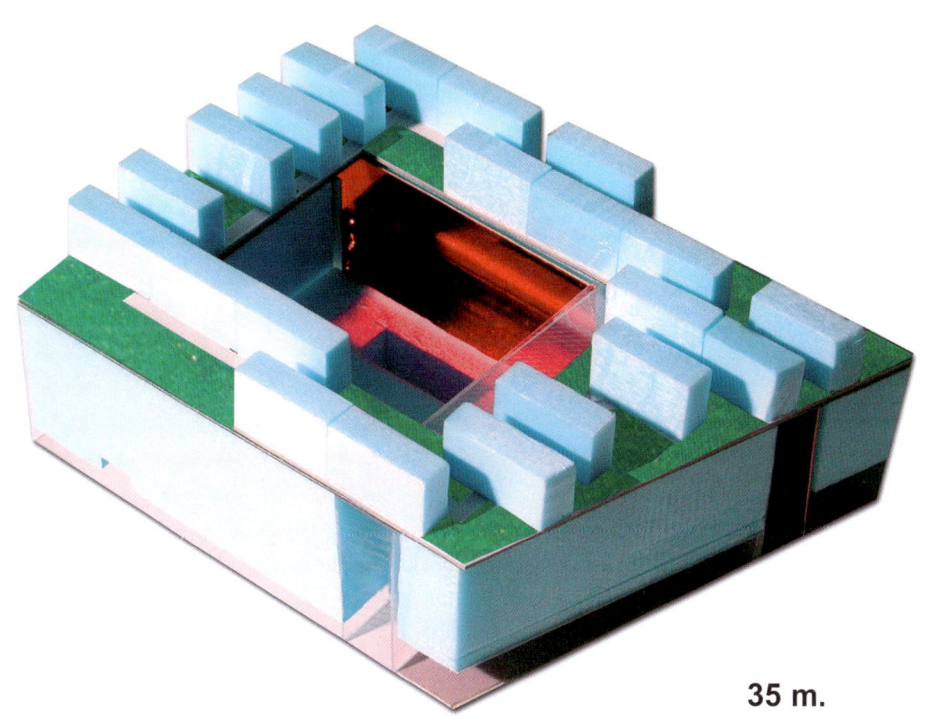

35 m.

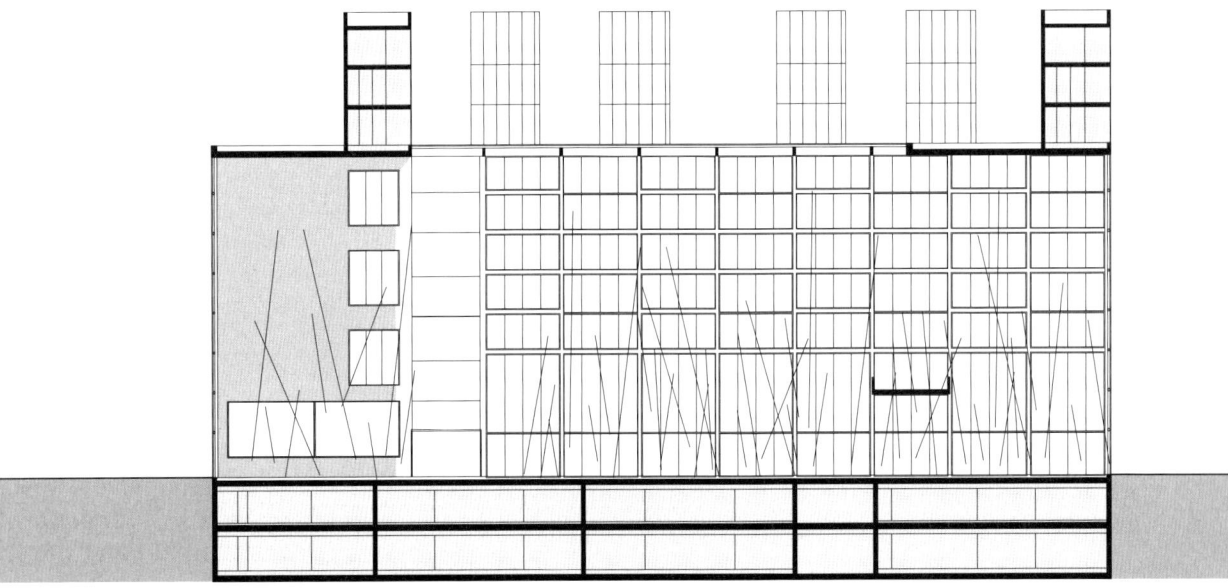

bamboo garden elevation, east side

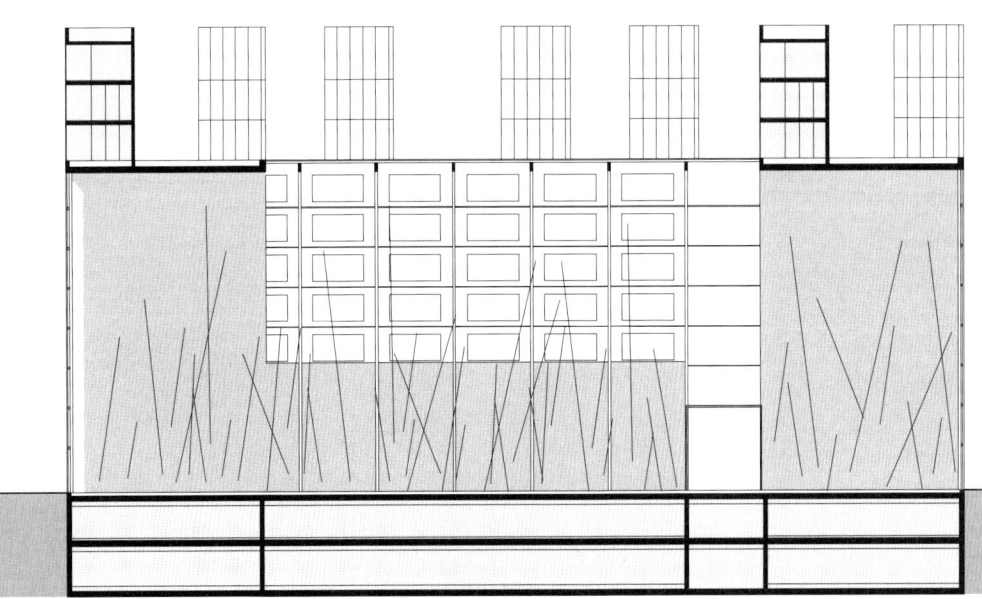

bamboo garden elevation, west side

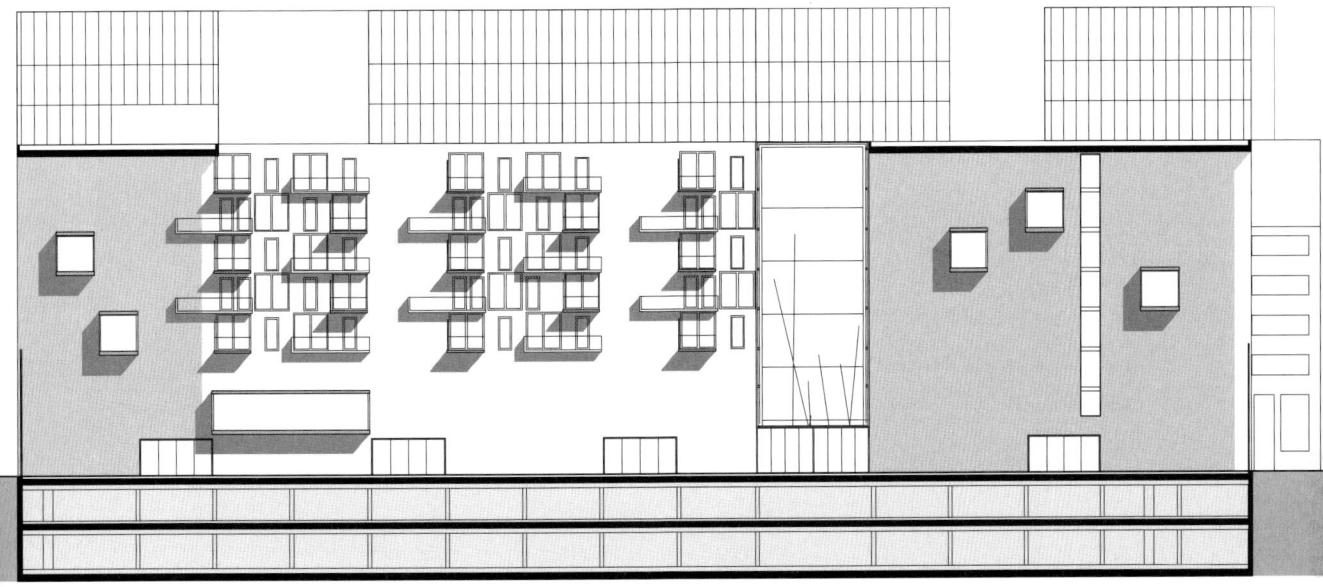

inner street elevation, north side

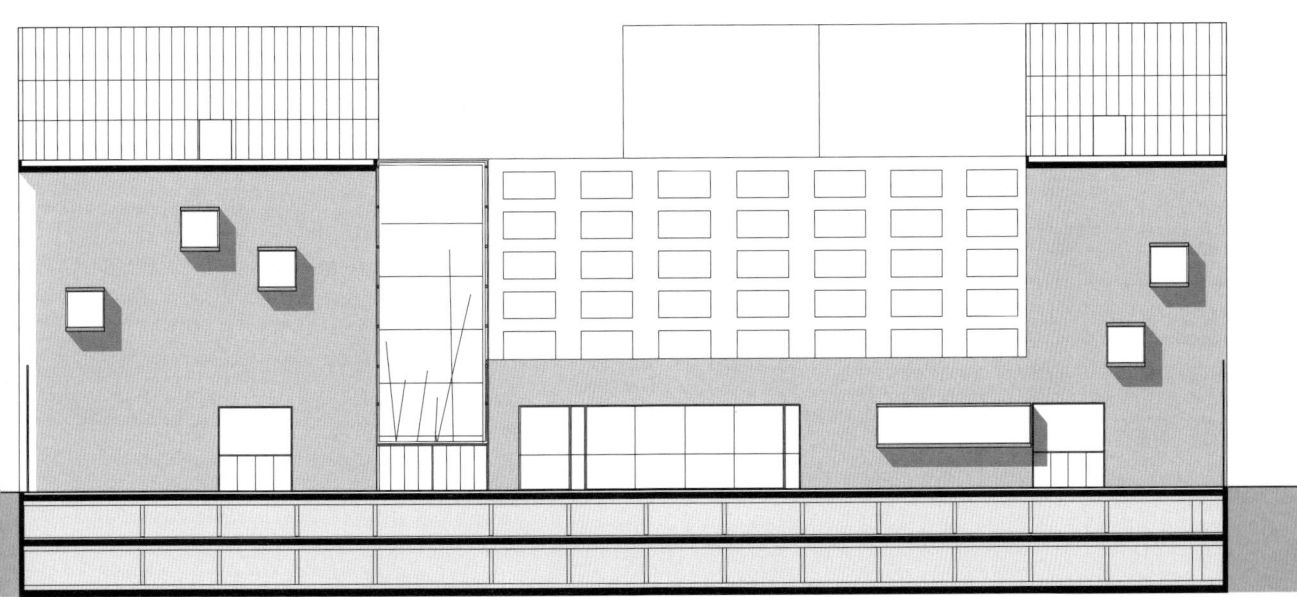

inner street elevation, south side

Oosterparkstraat
Time

For the dwellings in the Tweede Oosterparkstraat in Amsterdam, the nineteenth–century architecture and urban planning provided a basis for the development of the housing typology on the ground floor. The standard retail unit with a dwelling to the rear and upstairs is transformed into a maisonette with the dining room overlooking the street, while the living space on the garden side and upstairs is more private in character.

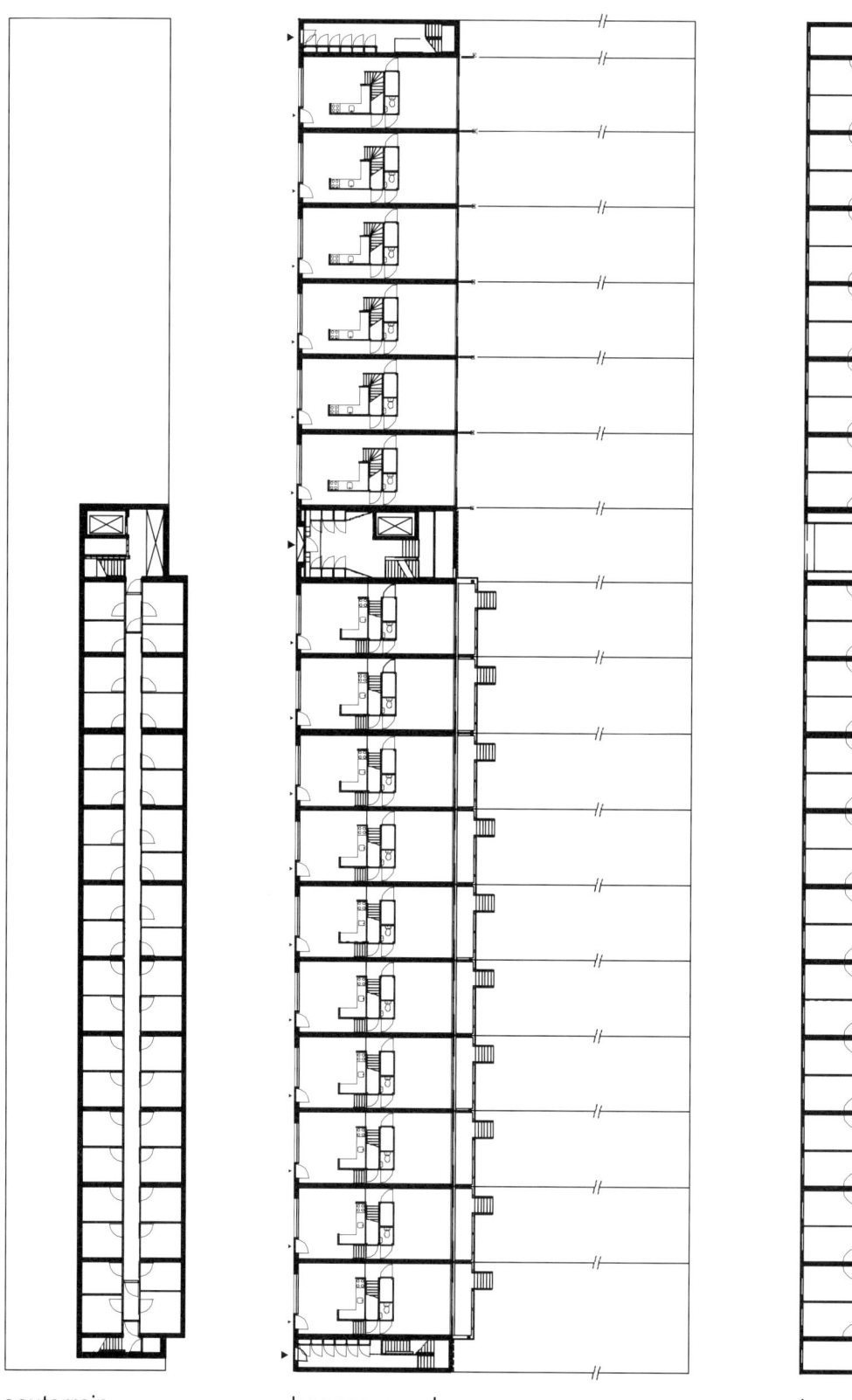

souterrain

begane grond

1e verdieping

0   10m

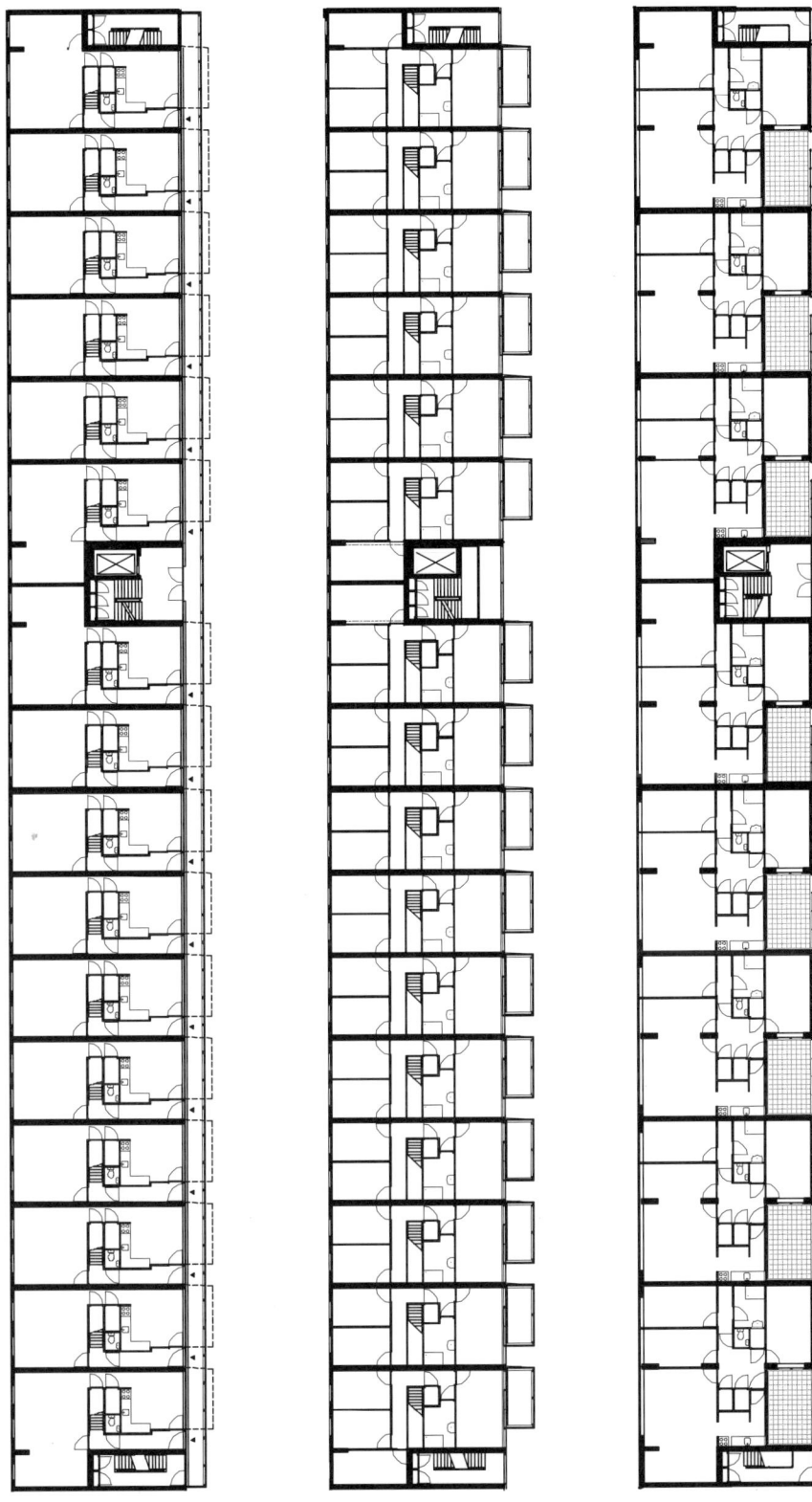

2e verdieping   3e verdieping   dak verdieping

Ouagadougou
Time

An urban plan was devised for an area of the capital of Burkina Faso based on the structure that has spontaneously developed, i.e. without planning. The paths in the area were widened, extended or diverted slightly, so that in principle each plot is accessible by car. Within a short time, the new network of paths stimulated all kinds of urban activities along the arterial streets, such as shops and businesses. It also presented an opportunity for the local authorities to lay water mains and sewers. Elaborating the pattern of streets that was already present opened the way for social and economic development.

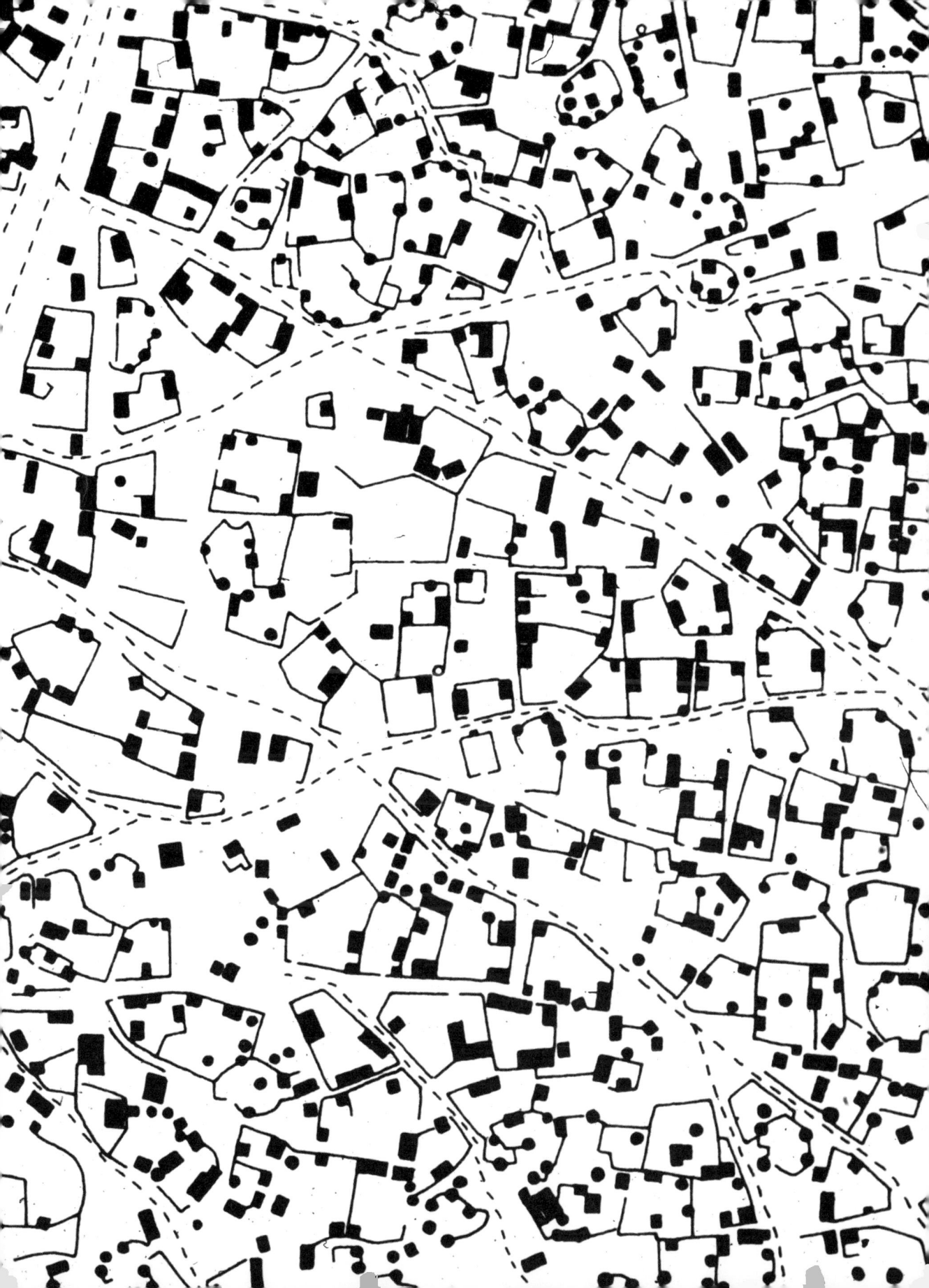

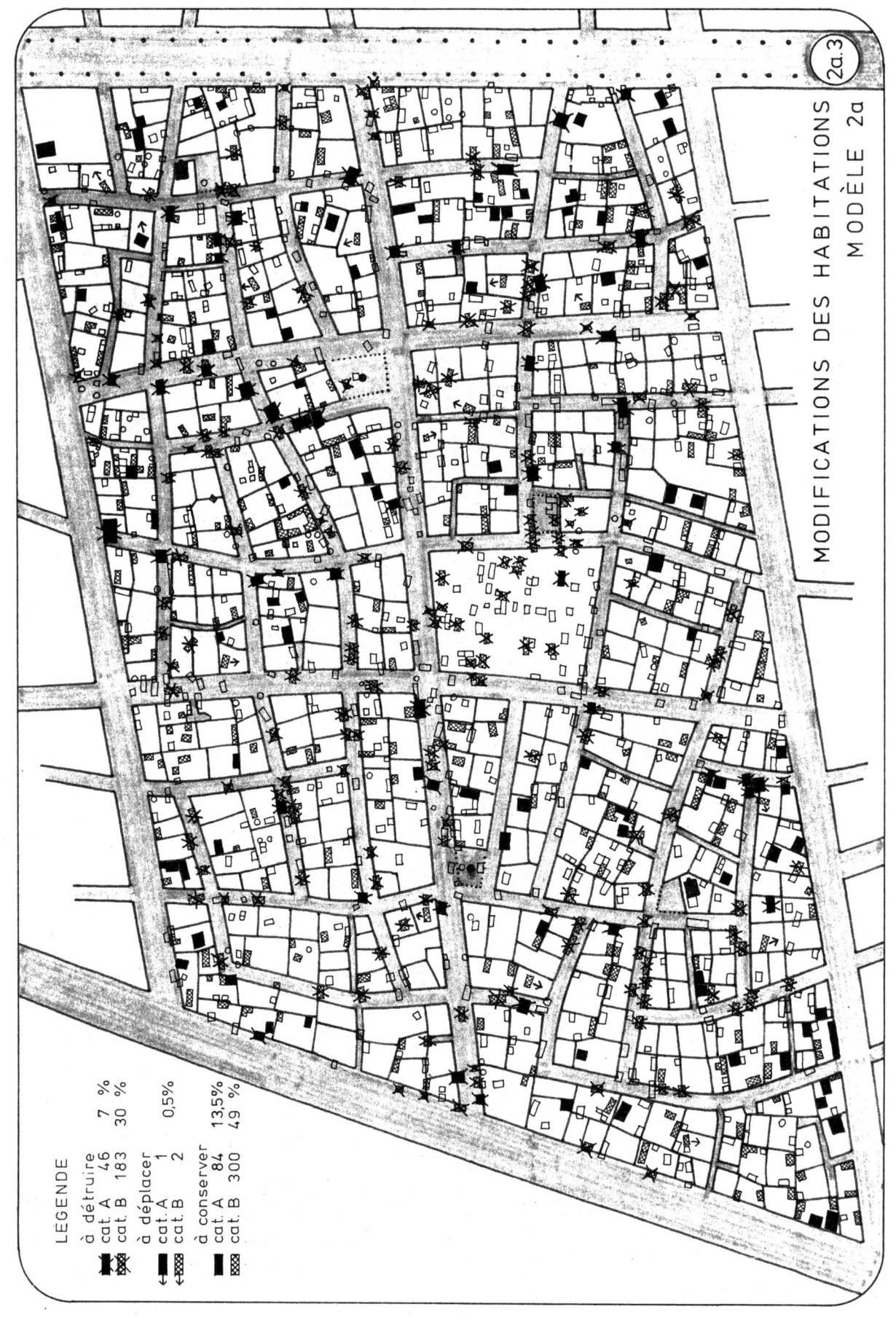

LÉGENDE

à détruire
cat. A   46   7 %
cat. B 183 30 %

à déplacer
cat. A   1
cat. B   2   0,5%

à conserver
cat. A   84 13,5%
cat. B 300 49 %

MODIFICATIONS DES HABITATIONS
MODÈLE 2a

2a.3

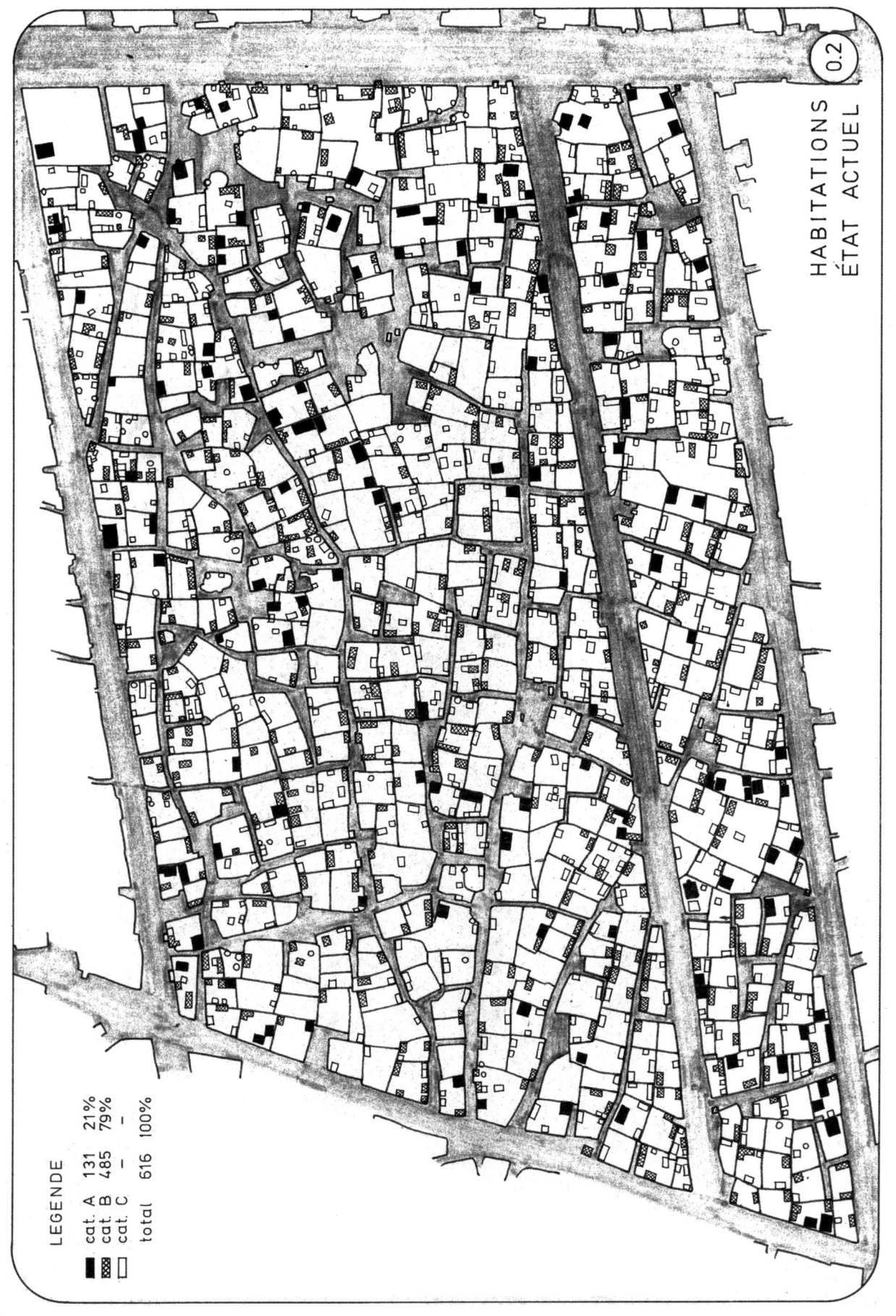

HABITATIONS
ÉTAT ACTUEL

0.2

LEGENDE

cat. A   131   21%
cat. B   485   79%
cat. C   -   -

total   616   100%

Nonnenveld
Time

The residential block stands on the edge of the Chassé Park in Breda, a former barracks for which OMA devised the urban plan. The building belongs as much with the old city as with the new. The building completes the urban block and is tailored to it in its various heights and in the continuation of the wall. At the same time it asserts its independence, maintaining a certain distance from the old city and turning towards the park.

kelder

begane grond

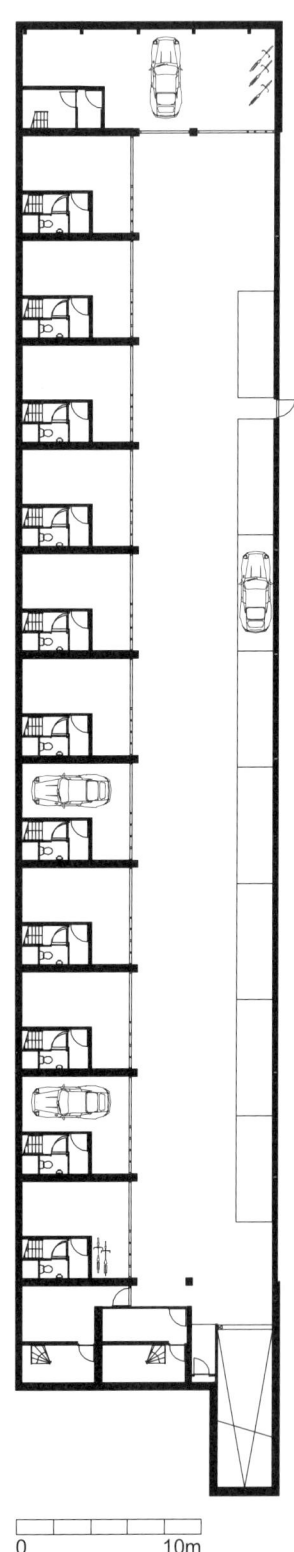
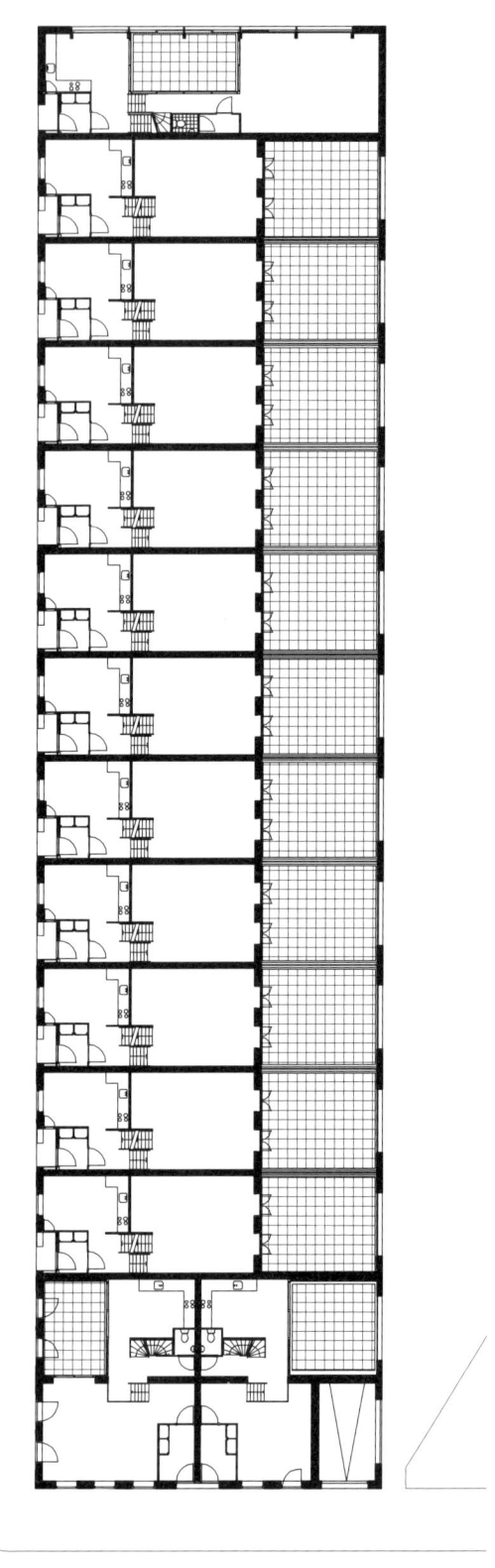

0   10m

1e verdieping

2e verdieping

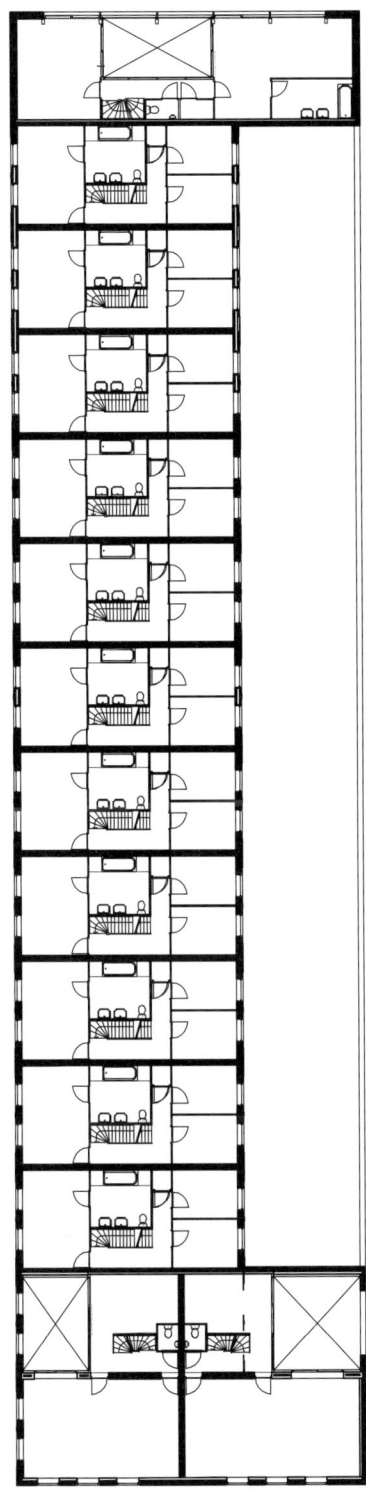
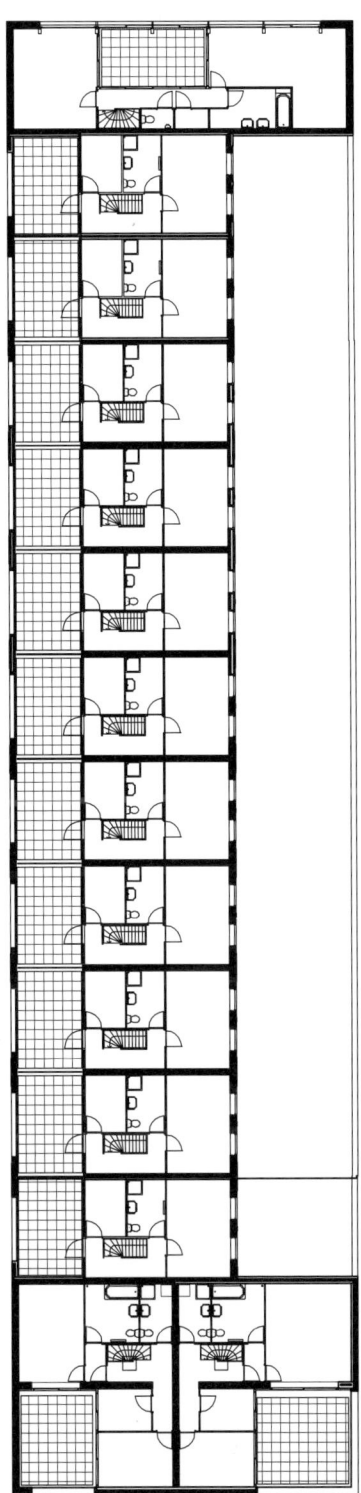

Erasmusgaarde
Time

Time is the most important theme in this project for the restructuring of Field 17 in the redevelopment of Zuidwest in The Hague. The project is a transformation of the structure of Zuidwest, which was originally based on an urban layout by W.M. Dudok that was elaborated by J.H. van den Broek. The architecture is also a transformation of what there was, retaining three crucial aspects of the original organization. • The height of the construction, 13 metres, has been retained. The proportion of the elevations that is open – 37 per cent – was also left unchanged. Since the dimensions beneath and alongside the fenestration are fixed at 40 centimetres and the height above the windows at 80 centimetres, the layout of the facade has been altered and varies depending on the height and width of the dwelling. • In addition, the grid of the facade has been shifted up 60 centimetres in each successive block, which creates an asymmetry in each successive facade, emphasizing the continuity around the entire block. The blocks no longer have a rear, and the elevations are of equal importance on every side. • Another important aspect of the transformation is the change in typology: from the original porch-access dwellings to units with front–door access to the street which means they resemble true houses.

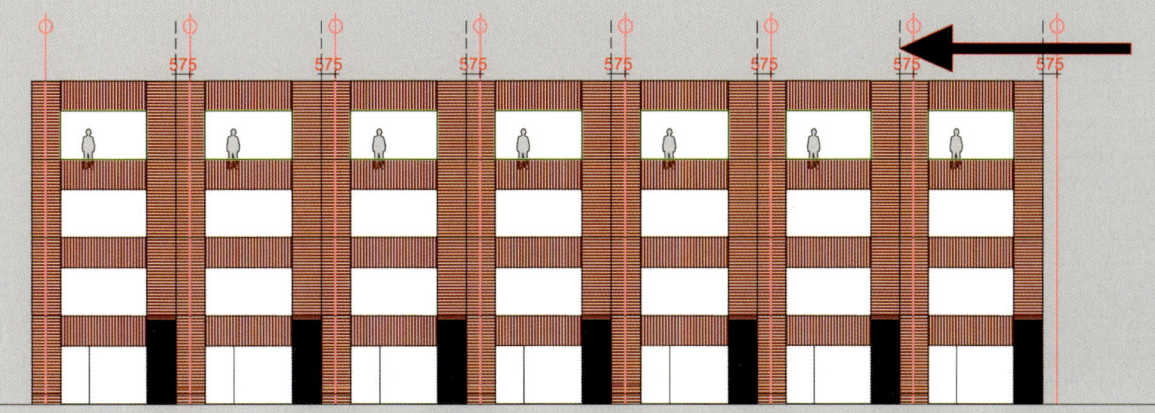

575   575   575   575   575   575   575

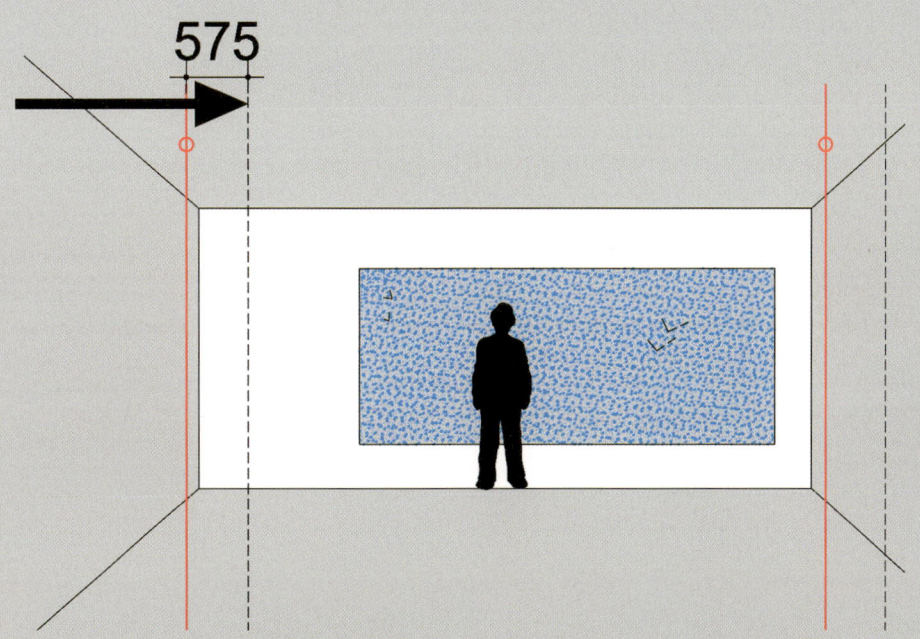

575

Interiority

The transition between inside and outside, between interior and exterior, is rarely unequivocal. What at first glance seems like a clear–cut transition between inside and outside does not always prove to be the case. Where something belongs or what it can be counted among depends on perception and interpretation, from the way in which something is looked at. A room and a whole city can both be read as an interior. This relativity leads to nuance in space and use.

Interiority

De Aker
Interiority

The patio houses in De Aker in Amsterdam can be understood as a succession of spaces, the character of which changes in gradations, from open and public to enclosed and private. These transitions from public via collective to private and from exterior to interior occur gradually, thus making it clear that each element in the series is linked with what comes immediately before and after, but without this occasioning a lack of clarity about the difference between each of the sequence of spaces in the series. Street and inner street, inner street and patio, patio and house – each step in the transition from what is outside and what is inside is consistently relatively small, no matter how great the sum difference between the outside world of the street and the interior worlds of the patio houses might be.

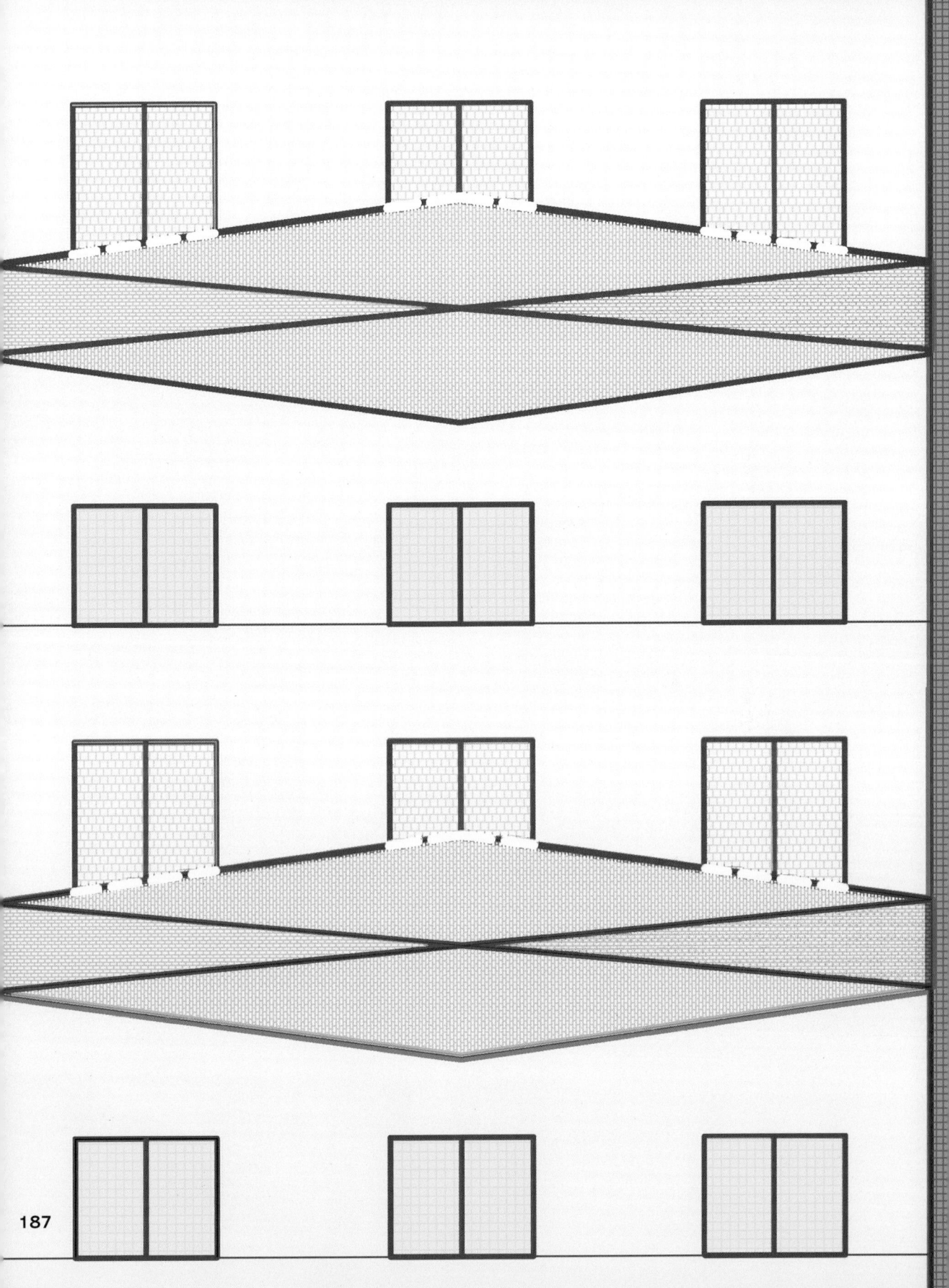

Ciboga
Interiority

The Ciboga district on the edge of Groningen's city centre is built up from 'floes', each expressly a different figure and together forming an area without streets that is completely unlike the city surrounding it. Floes 3, 4 and 5 all have a different form and are built using different materials. Floe 3 is rounded off as a distribution point, Floe 4 marks the point of access from the city and is folded around the existing construction, and Floe 5 has a serrated form. • Within the context of the city, the Ciboga area as a whole forms a distinctive space, and each of the floes has a different character. In Floe 5 there is yet another distinctive space, in the form of a communal courtyard garden, surrounded by houses which have patios on the ground floor and rooftop terraces on the uppermost floor.

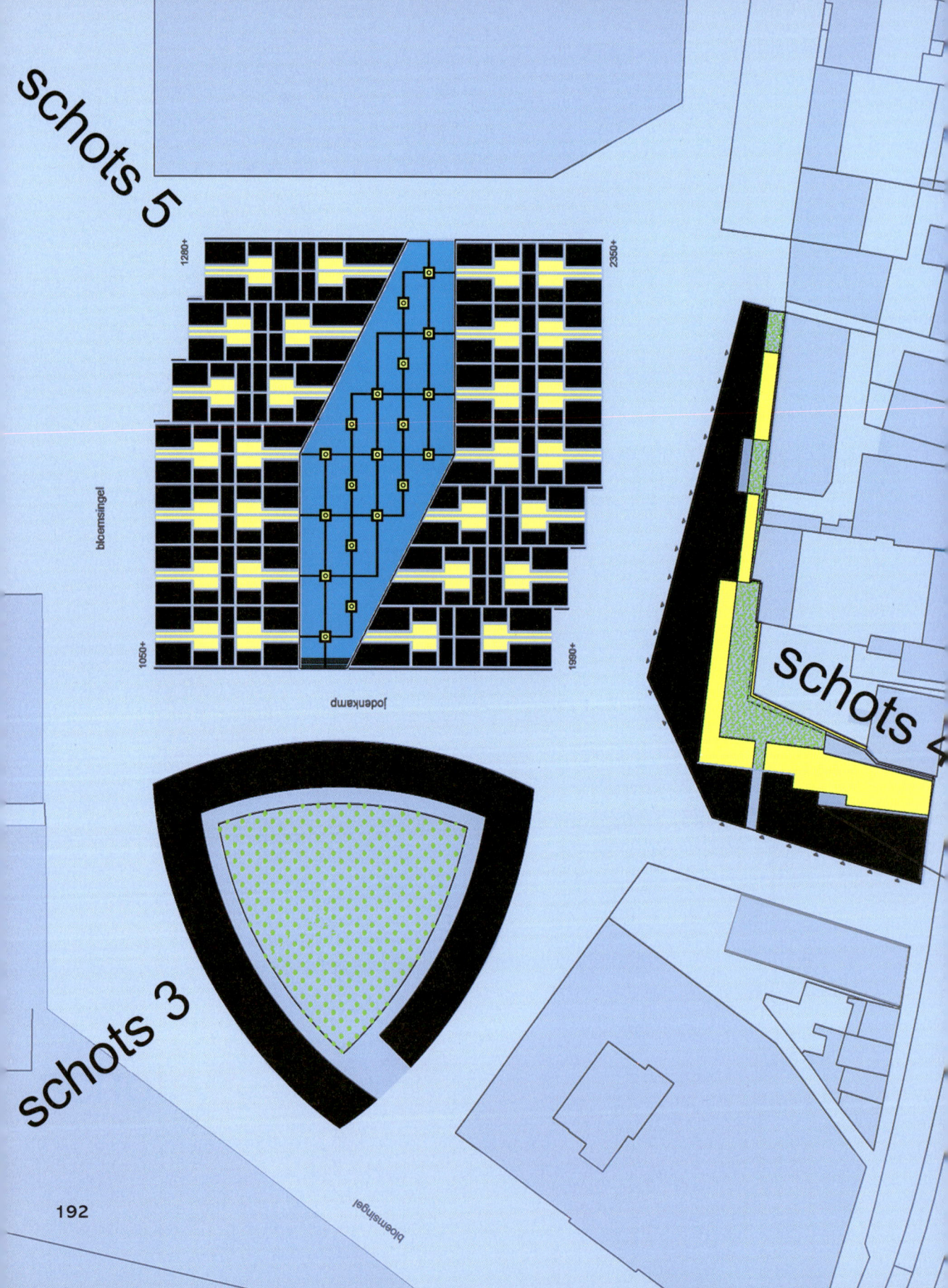

schots 5

schots 3

schots 4

1280+

2350+

1050+

1990+

bloemsingel

jodenkamp

bloemsingel

Haarlem
Interiority

The front elevation of this 250–metre–long strip with two residential blocks in the Zuiderpolder in Haarlem consists of 21–metre–wide upper dwellings, each extending across three maisonettes beneath. In front of this strip there is a thin arcade. The series of ultra–slender columns demarcates a zone that belongs to the street as well as the dwellings and rooftop terraces, to both the public space and the building.

Witbrant
Interiority

The essence of the urban plan by Jacq. de Brouwer for the Witbrant–Oost development in Tilburg is the spatial sequence of wooded landscape, neighbourhood, house and patio. This sequence of exterior and interior space is amplified inside the dwellings, especially those that extend across the full depth of the blocks. These dwellings (and their inhabitants) therefore forfeit a singular, introverted orientation towards the patio, but there is also constant contact with a subsequent exterior space, the street, the landscape.

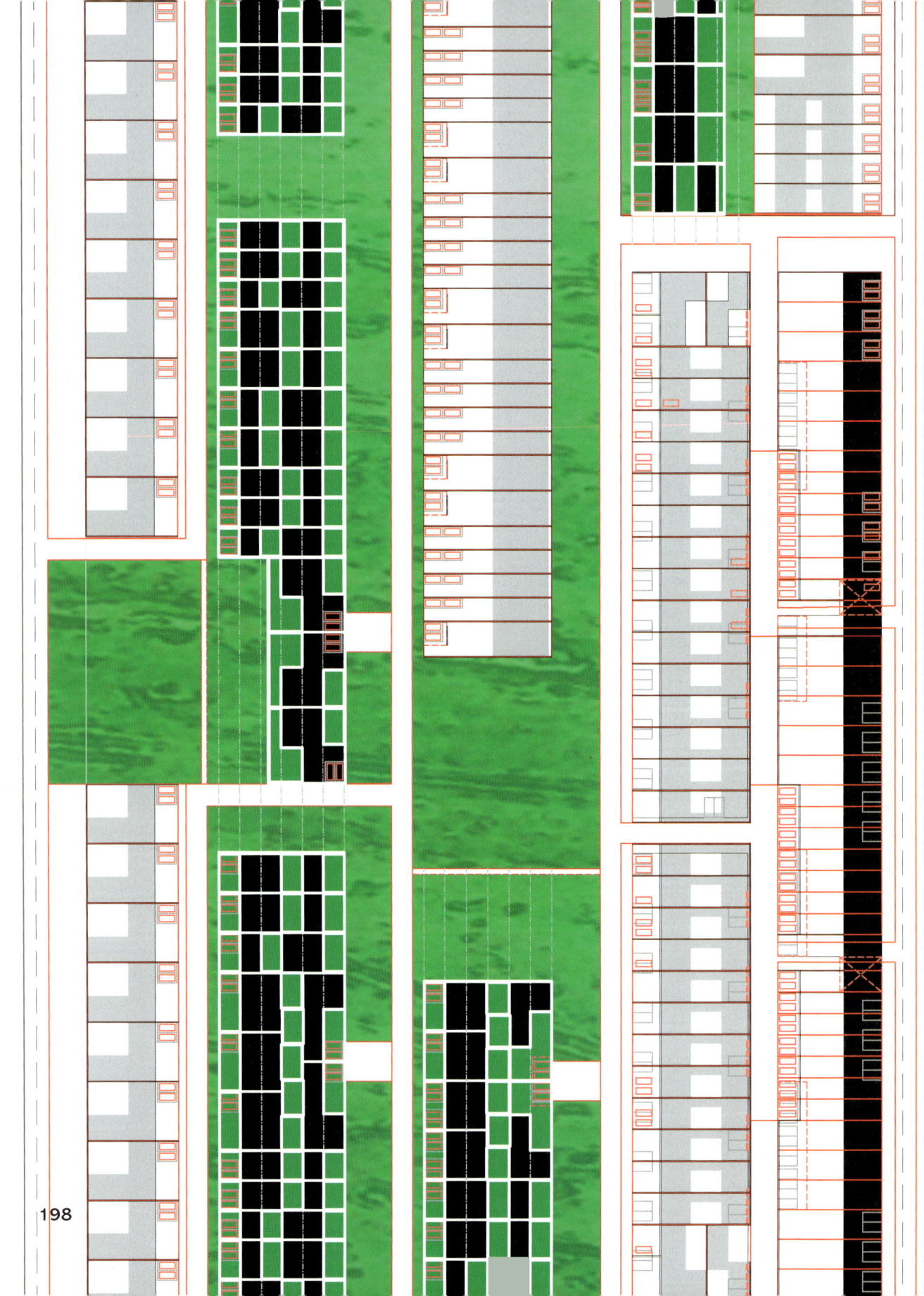

Sporenburg
Interiority

The plan by West 8 for the Borneo and Sporenburg peninsula in Amsterdam's Eastern Harbour District is dictated by the mass and the open-work space of street, car port, atrium and rooftop terrace. At the ends of the peninsula, where the linear rows of dwellings terminate, it is not the mass but space that is the priority. The dwellings in the end blocks introduce a new kind of space in the area, also in terms of scale: the collective in–between space. This space was created by turning the terminations away from the extended strips. It is an interstitial space in two respects: between the row and termination, but also between the public street and the private space of the dwelling. This in–between space for collective use is visible from the street but closed off from it by a gate fence. This space is a component of the view from the dwellings, yet also filters the private view of the street and the water.

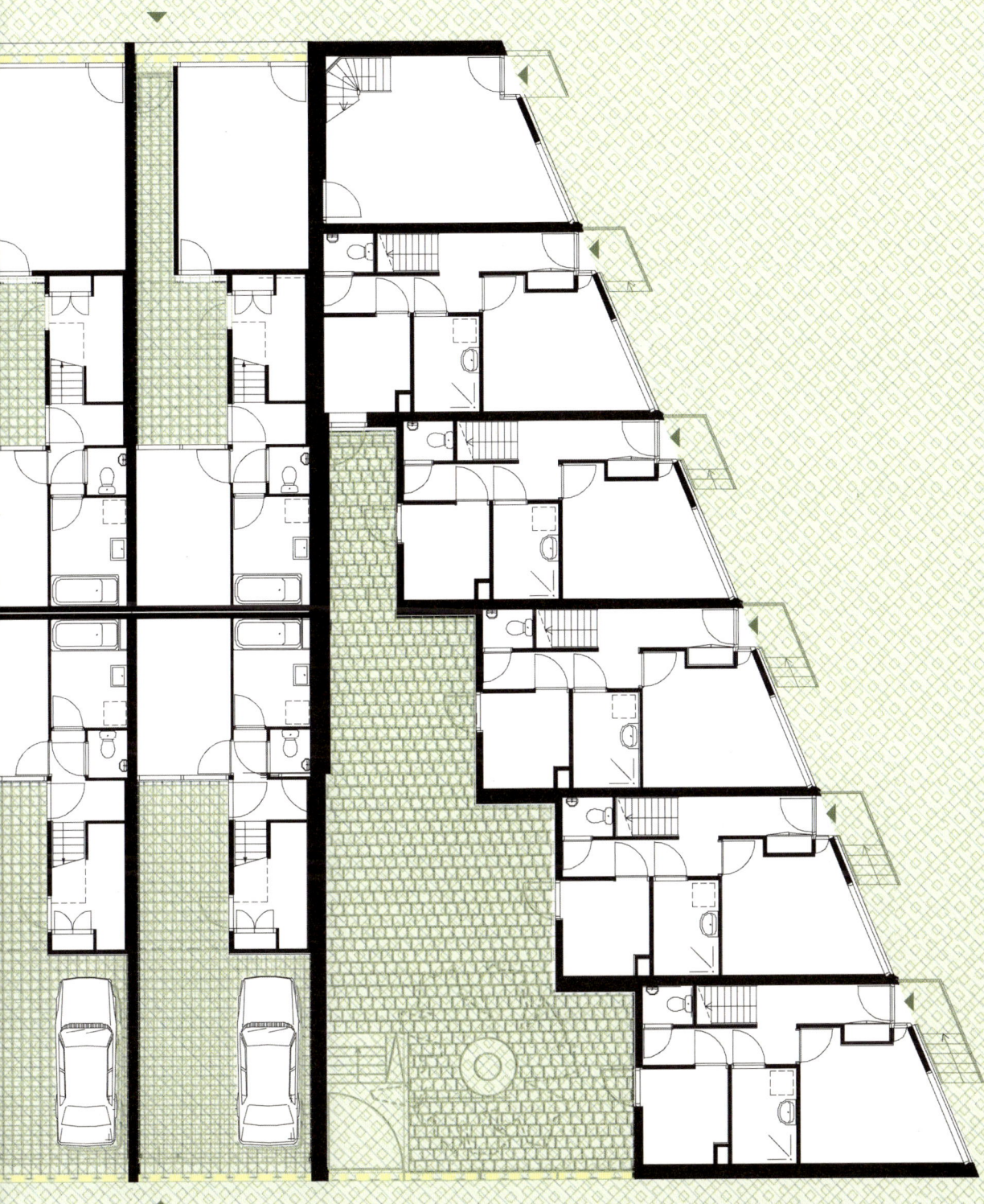

203

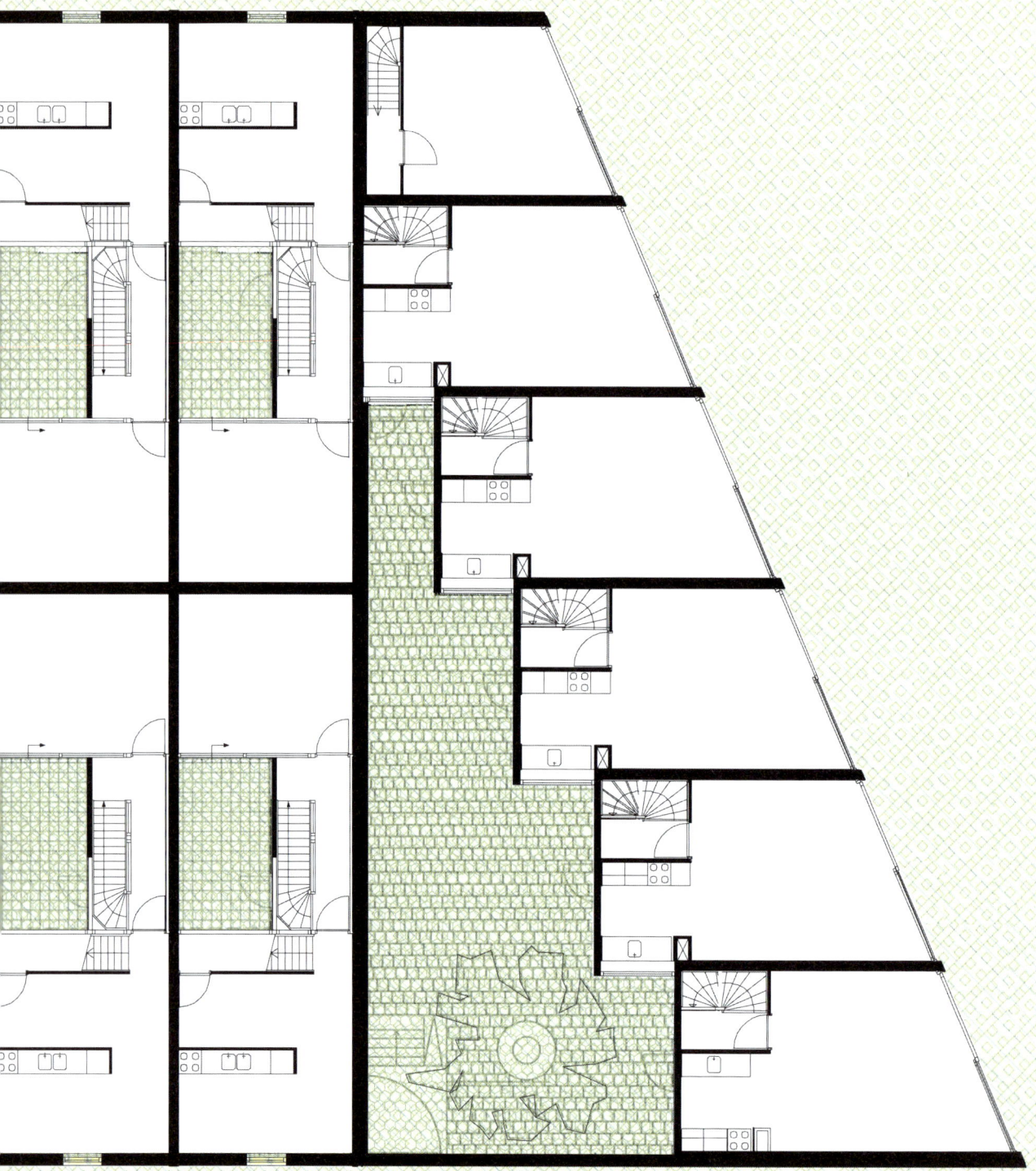

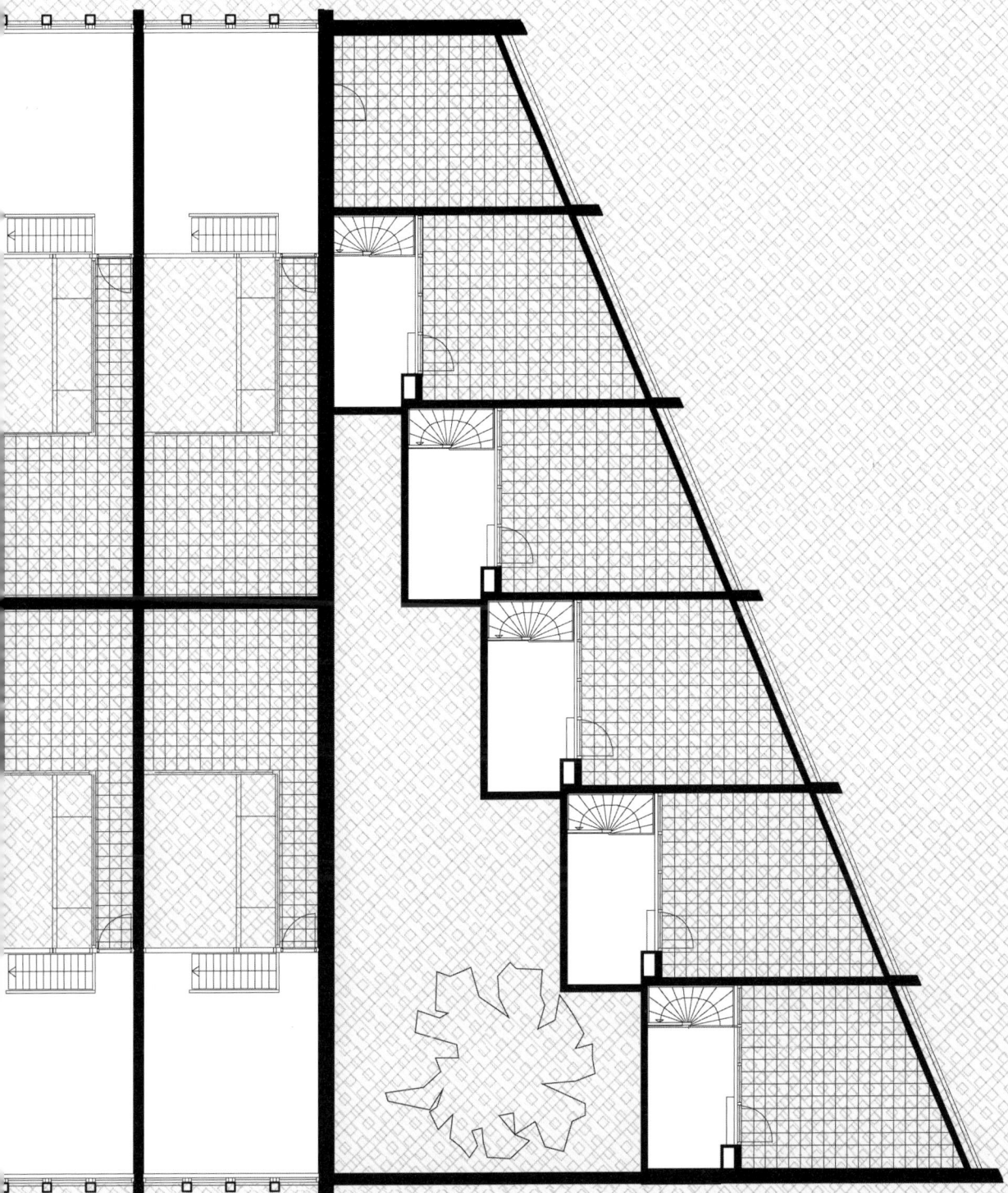

Ichthus
Interiority

The two orientations in the complex next to the Laurenskerk in Rotterdam are based on the cardinal orientations in the surrounding area. The cross that is formed by the 25–metre–high bamboo garden and the inner street – at the same height – is the mainstay of the plan. Both spaces are publicly accessible and belong to both building and city.

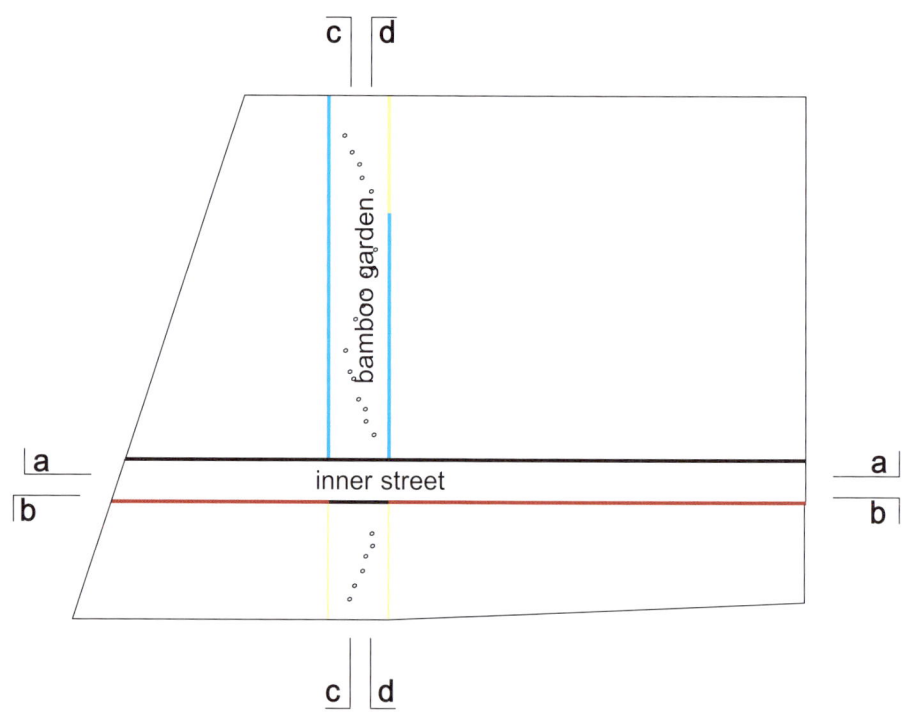

a
b

c d

bamboo garden

inner street

a
b

c d

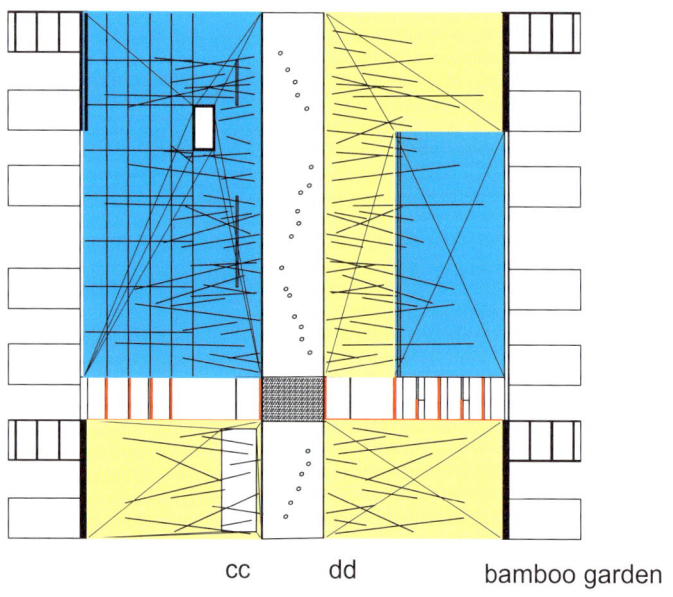

cc   dd   bamboo garden

208

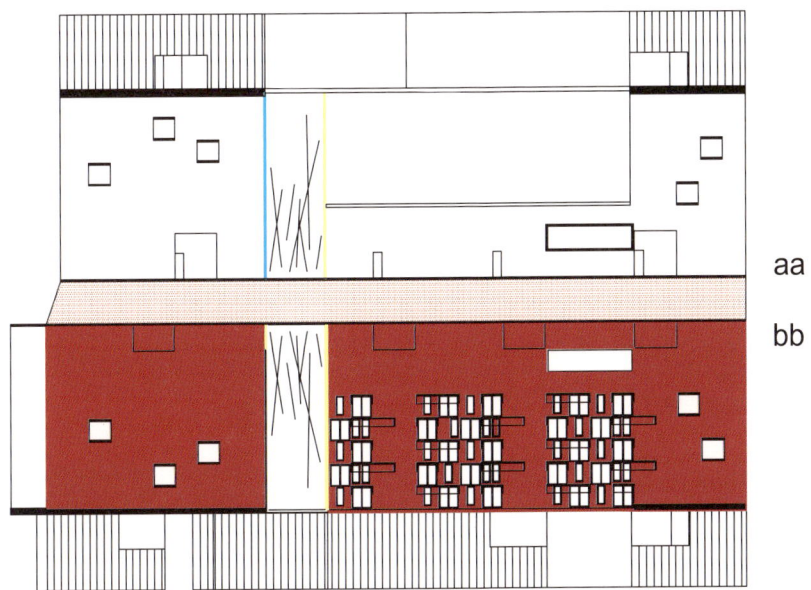

aa

bb

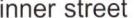

inner street

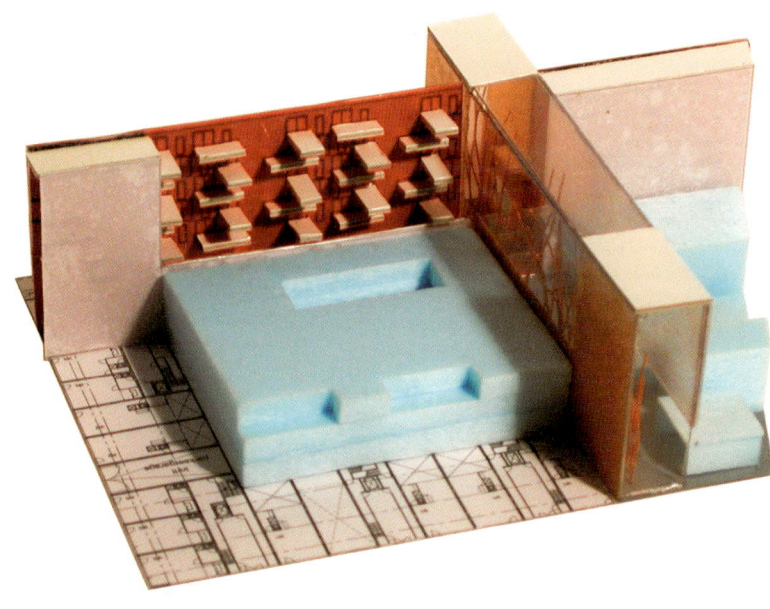

Seriality

With projects on a bigger scale, architecture often consists of the repetition of elements. Repetition is uninteresting in and of itself; how something is repeated and in what sequence and patterns is intriguing. The complexity or simplicity of this, which makes serial progressions from the repetition, can be used to organize architecture. In my work there are two kinds of seriality: in the combination of elements and in the overlapping of patterns.

Seriality

Bridges
Seriality

The design of bridges in the Amsterdam expansion district of IJburg is determined by the trusses—embankments—abutments series. For all the bridges the trusses are repeated, all sharing the same radius but varying in length depending on the specific situation. How many parallel trusses there are depends on the width of the bridge deck. The trusses and the various kinds of embankment and abutment, which might vary at the different ends of a bridge, results in each of them displaying a specific configuration and look.

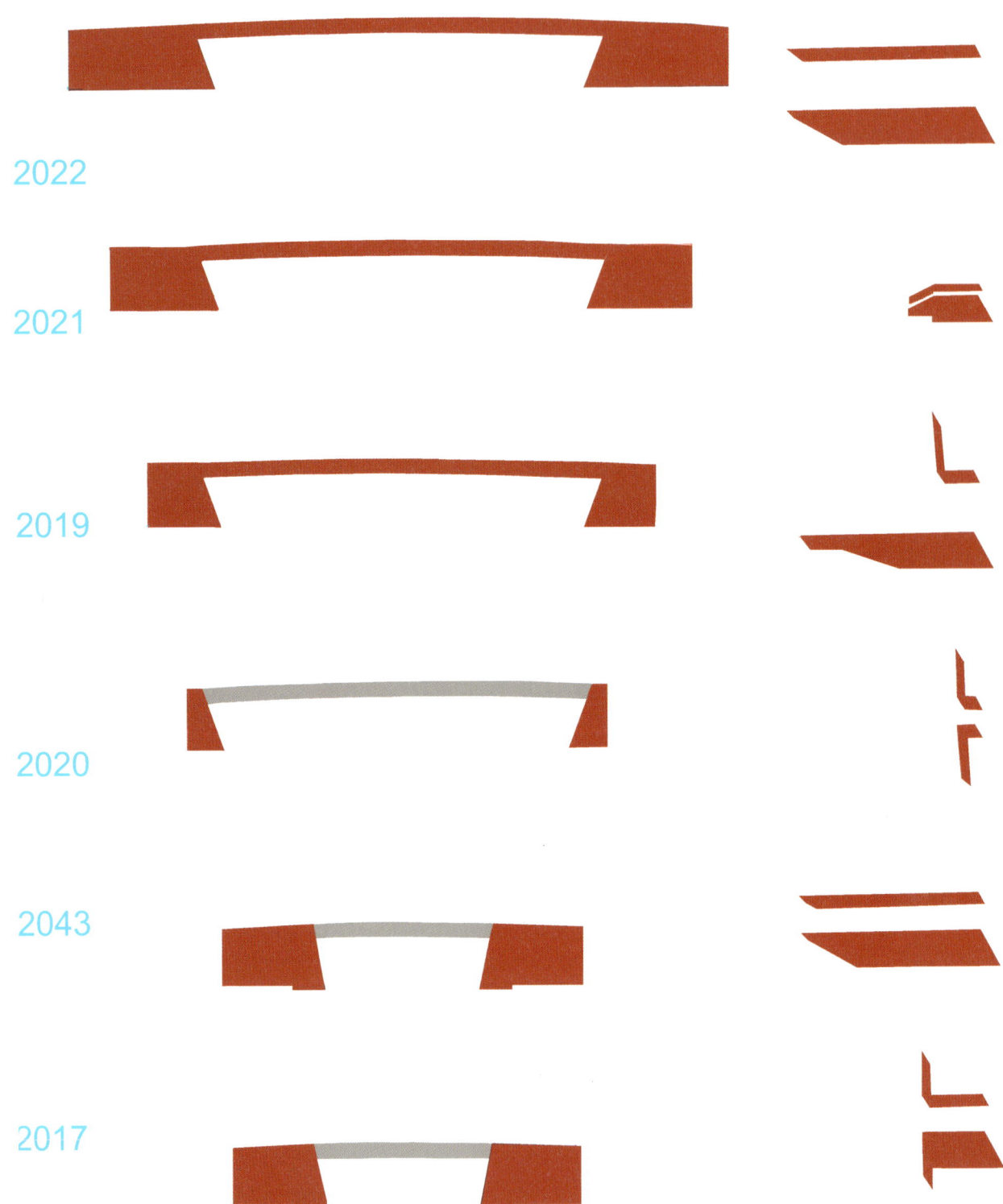

2022

2021

2019

2020

2043

2017

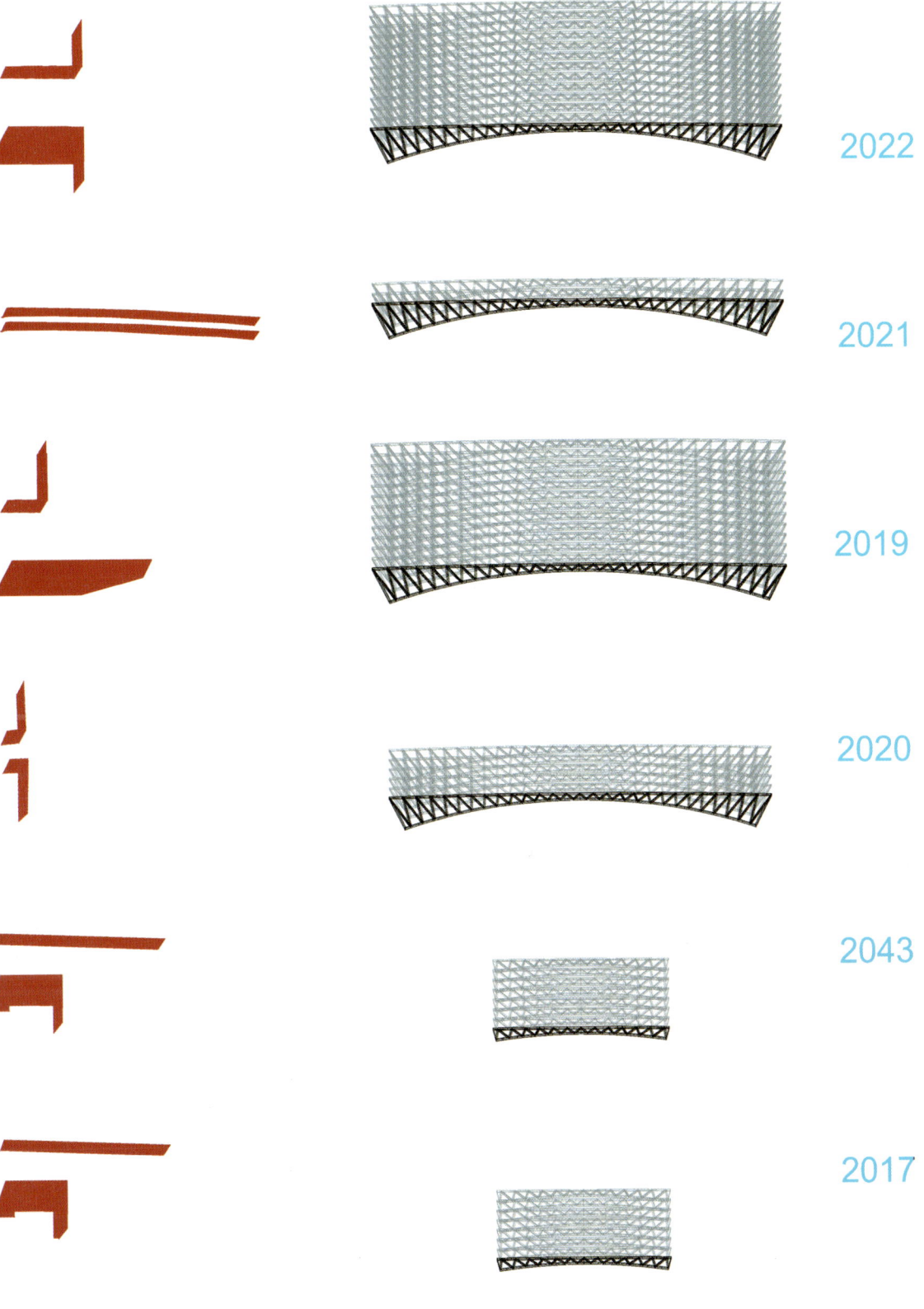

2022

2021

2019

2020

2043

2017

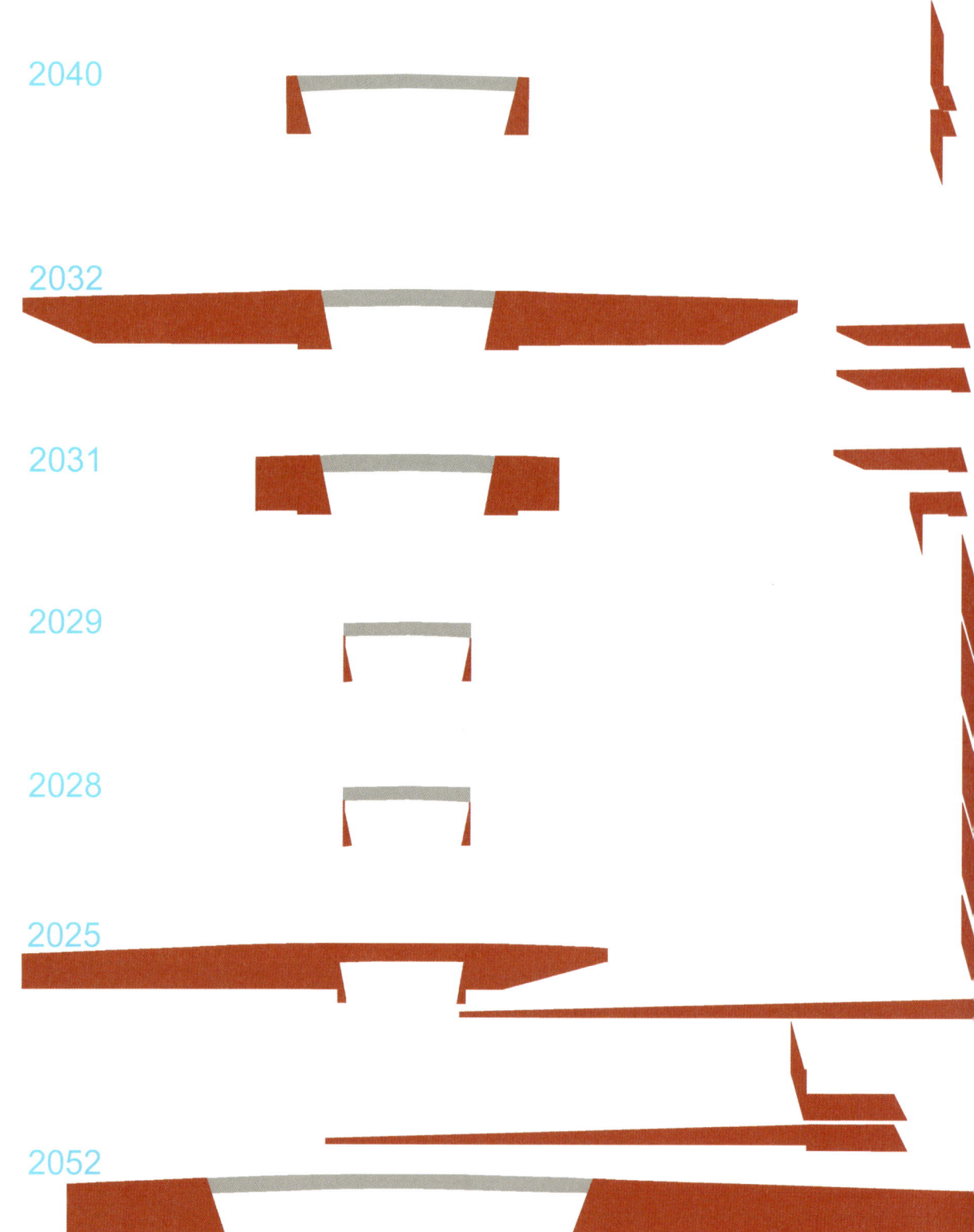

2040

2032

2031

2029

2028

2025

2052

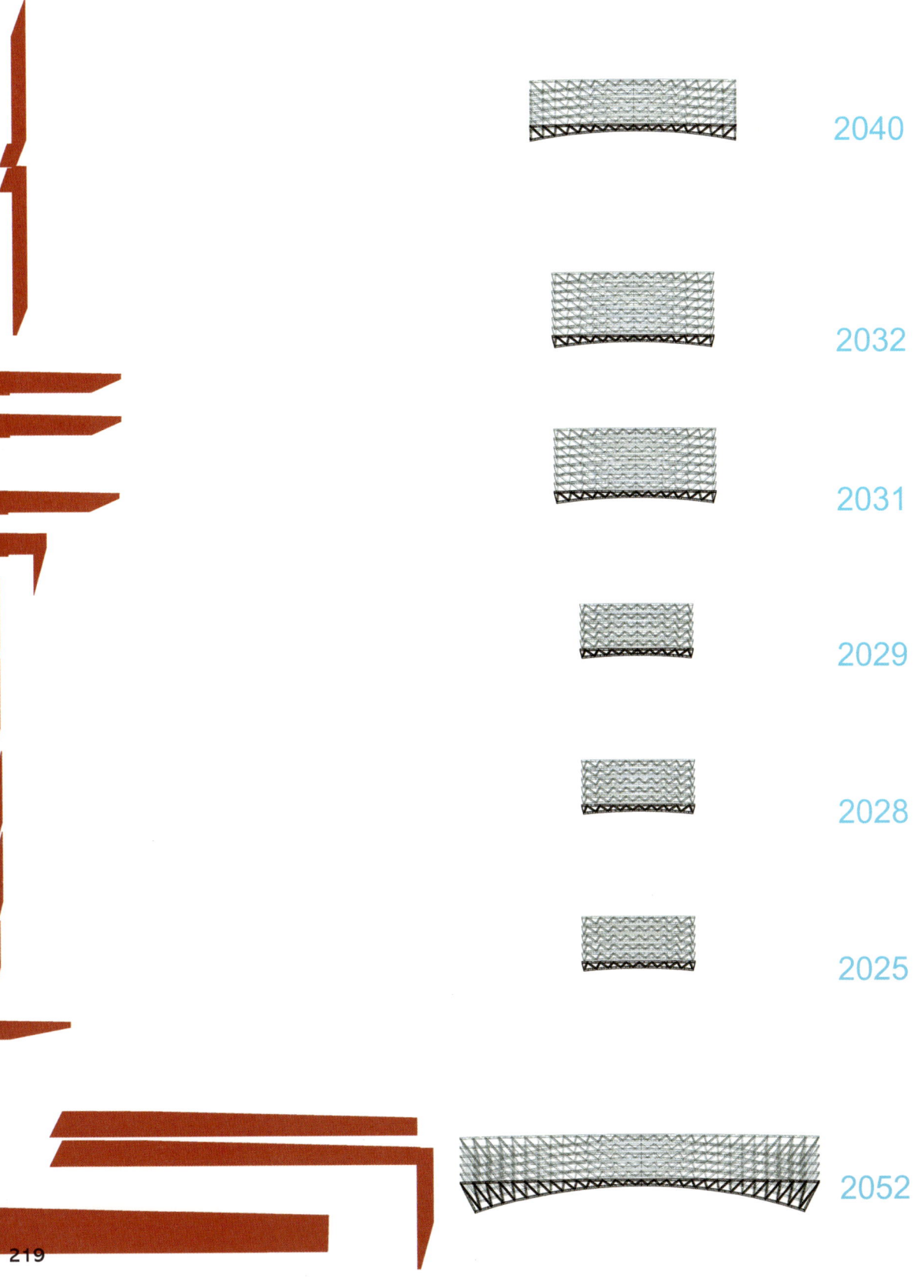

2040

2032

2031

2029

2028

2025

2052

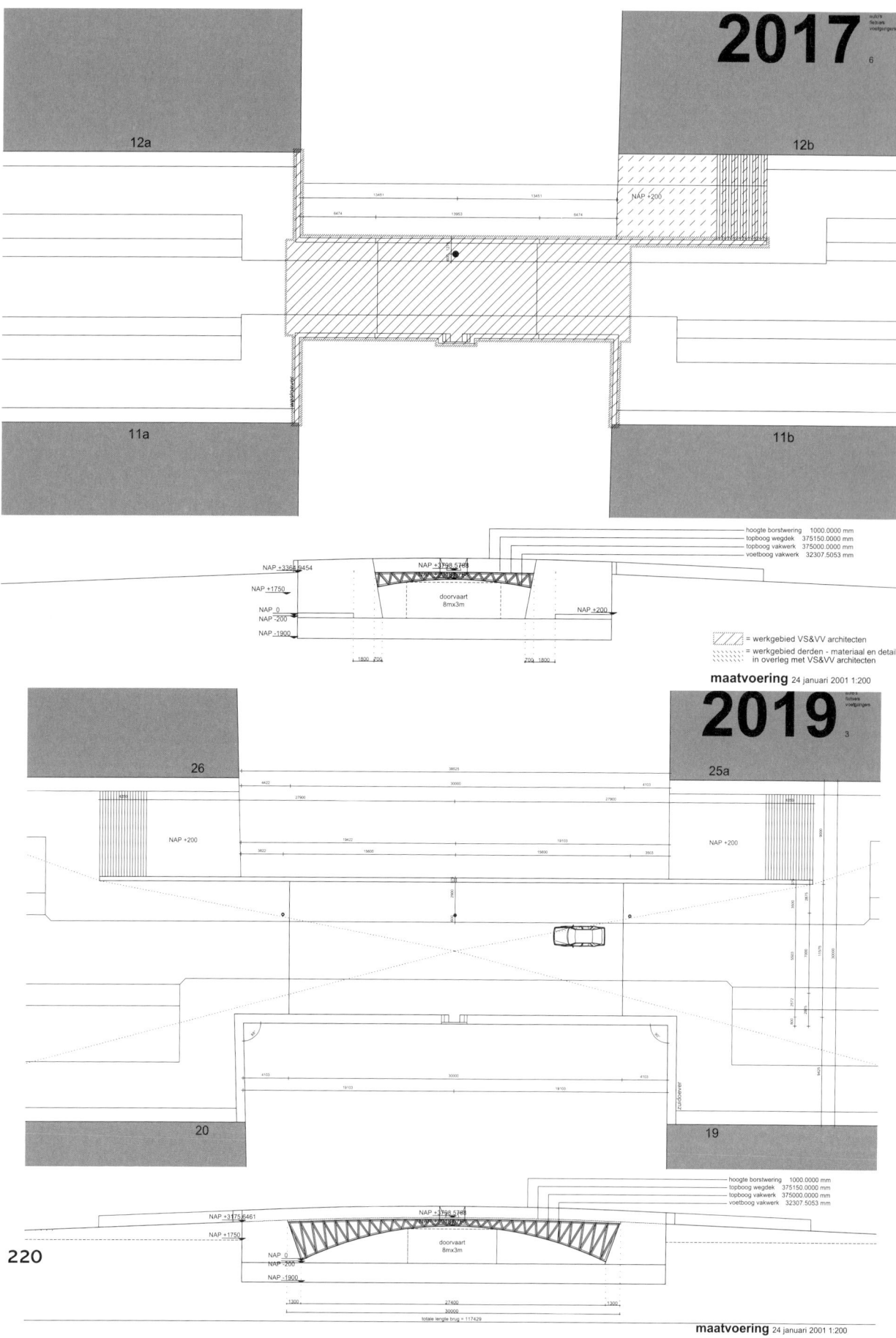

maatvoering 24 januari 2001 1:200

2019

maatvoering 24 januari 2001 1:200

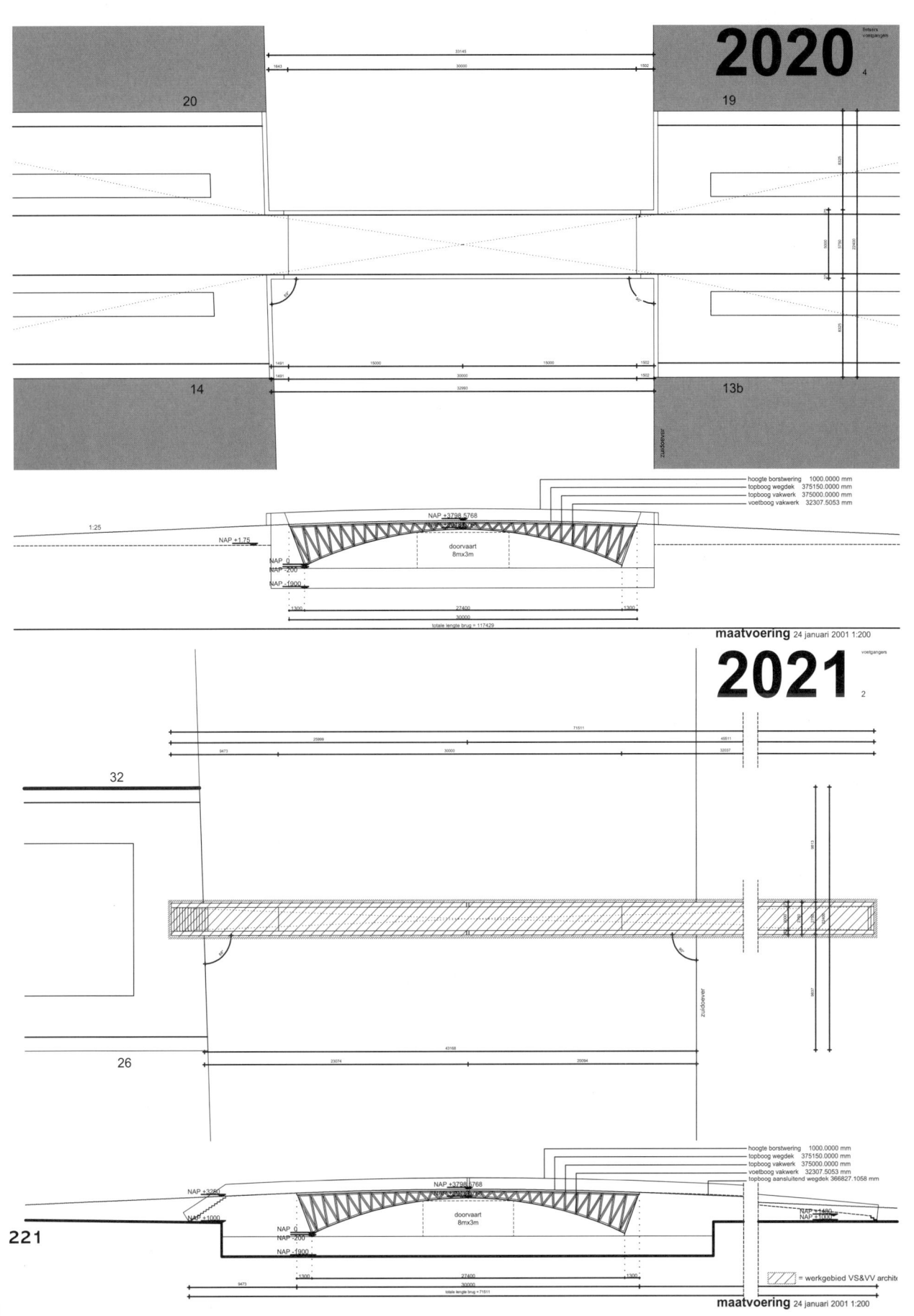

2020

maatvoering 24 januari 2001 1:200

2021

221

maatvoering 24 januari 2001 1:200

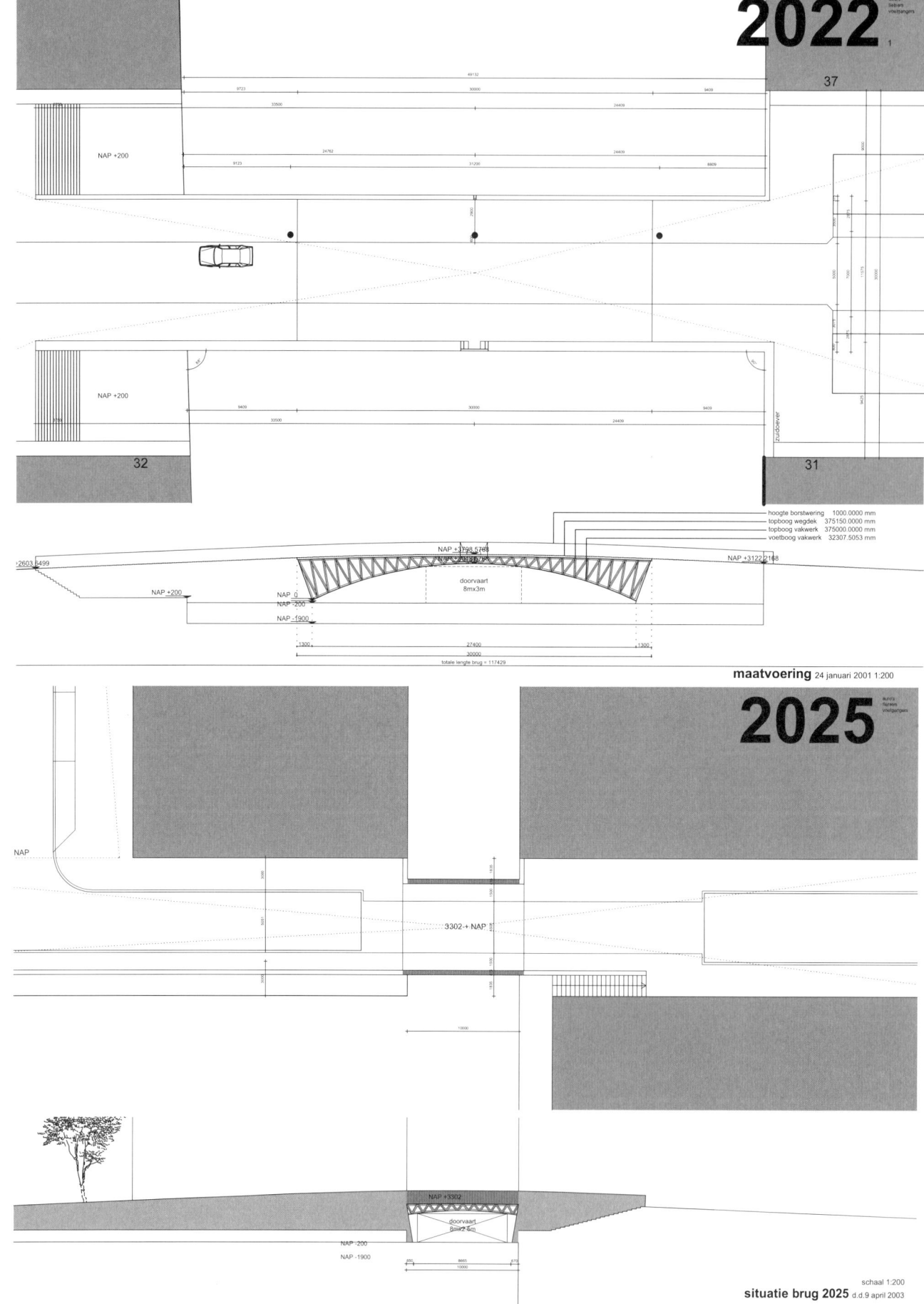

37

NAP +200

NAP +200

32

31

zuidoever

hoogte borstwering 1000.0000 mm
topboog wegdek 375150.0000 mm
topboog vakwerk 375000.0000 mm
voetboog vakwerk 32307.5053 mm

NAP +3798.5769
NAP +3122.2168

NAP +200
NAP 0
NAP -200
NAP -1900

doorvaart
8mx3m

1300   27400   1300
30000
totale lengte brug = 117429

maatvoering 24 januari 2001 1:200

NAP

3302 -+ NAP

10000

NAP +3302

doorvaart
8mx2.5m

NAP -200
NAP -1900

10000

schaal 1:200
situatie brug 2025 d.d.9 april 2003

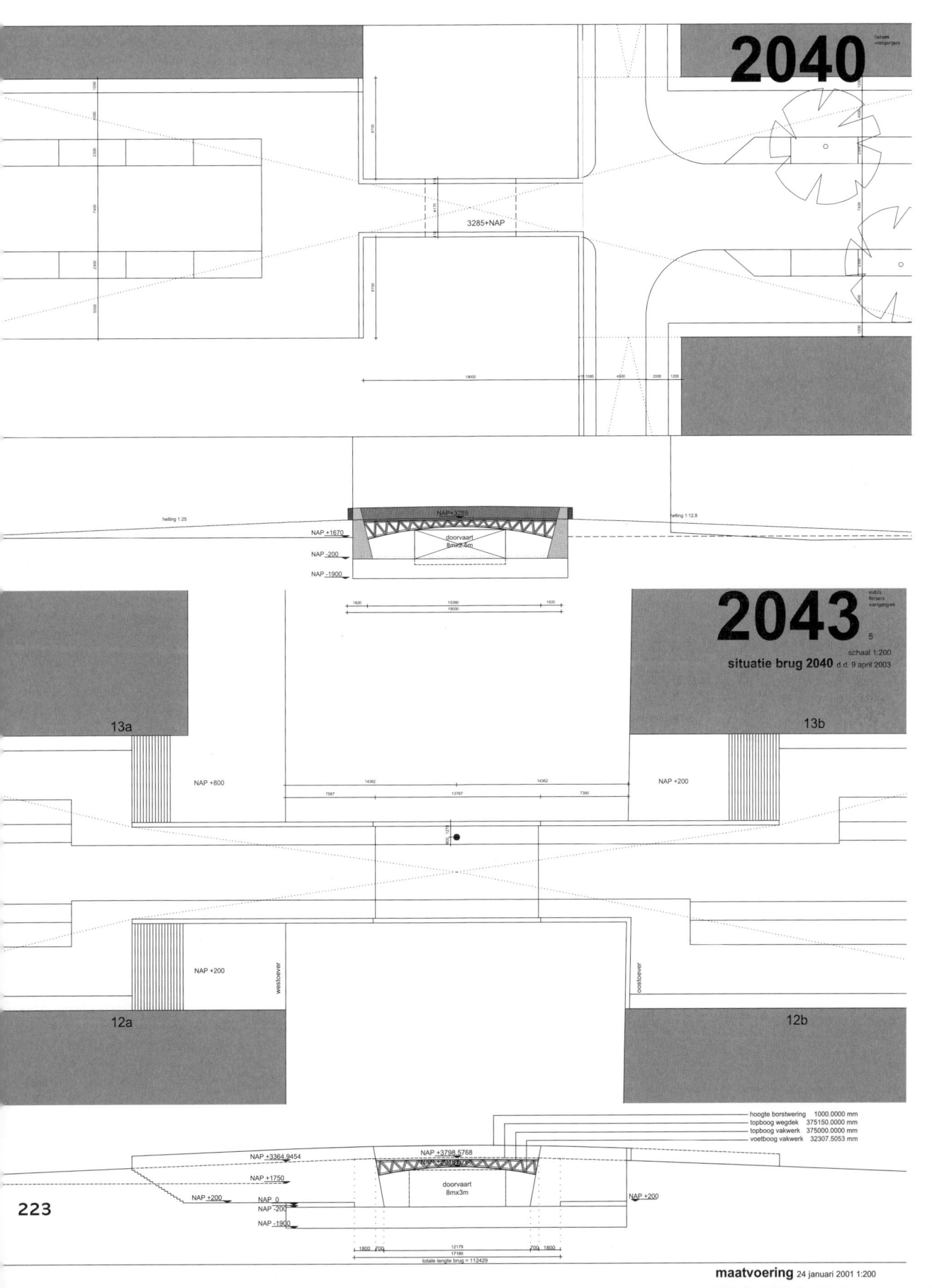

2040
fietsers
voetgangers

3285+NAP

helling 1:25   NAP+3285   helling 1:12,8

NAP +1670

NAP -200   doorvaart
8mx2,6m

NAP -1900

2043 5
auto's
fietsers
voetgangers

schaal 1:200
situatie brug 2040 d.d. 9 april 2003

13a   13b

NAP +800   NAP +200

westoever   oostoever

12a   12b

hoogte borstwering   1000.0000 mm
topboog wegdek   375150.0000 mm
topboog vakwerk   375000.0000 mm
voetboog vakwerk   32307.5053 mm

NAP +3364.9454   NAP +3798.5768

NAP +1750

NAP +200   doorvaart   NAP +200
NAP 0   8mx3m
NAP -200

NAP -1900

totale lengte brug = 112429

maatvoering 24 januari 2001 1:200

Ciboga
Seriality

The Ciboga area on the edge of Groningen's city centre is composed of 'floes'. The facade of Floe 4, which was folded around the existing architecture, is composed of glazed walls and facade elements of stainless steel or glass. The symmetrical constant is the deep–set fenestration; the asymmetrical constant is formed by the stainless steel panels surrounding the fenestration, which follow the same pattern for each house and alternate with the surface–embellished glass facade elements that are set in various positions.

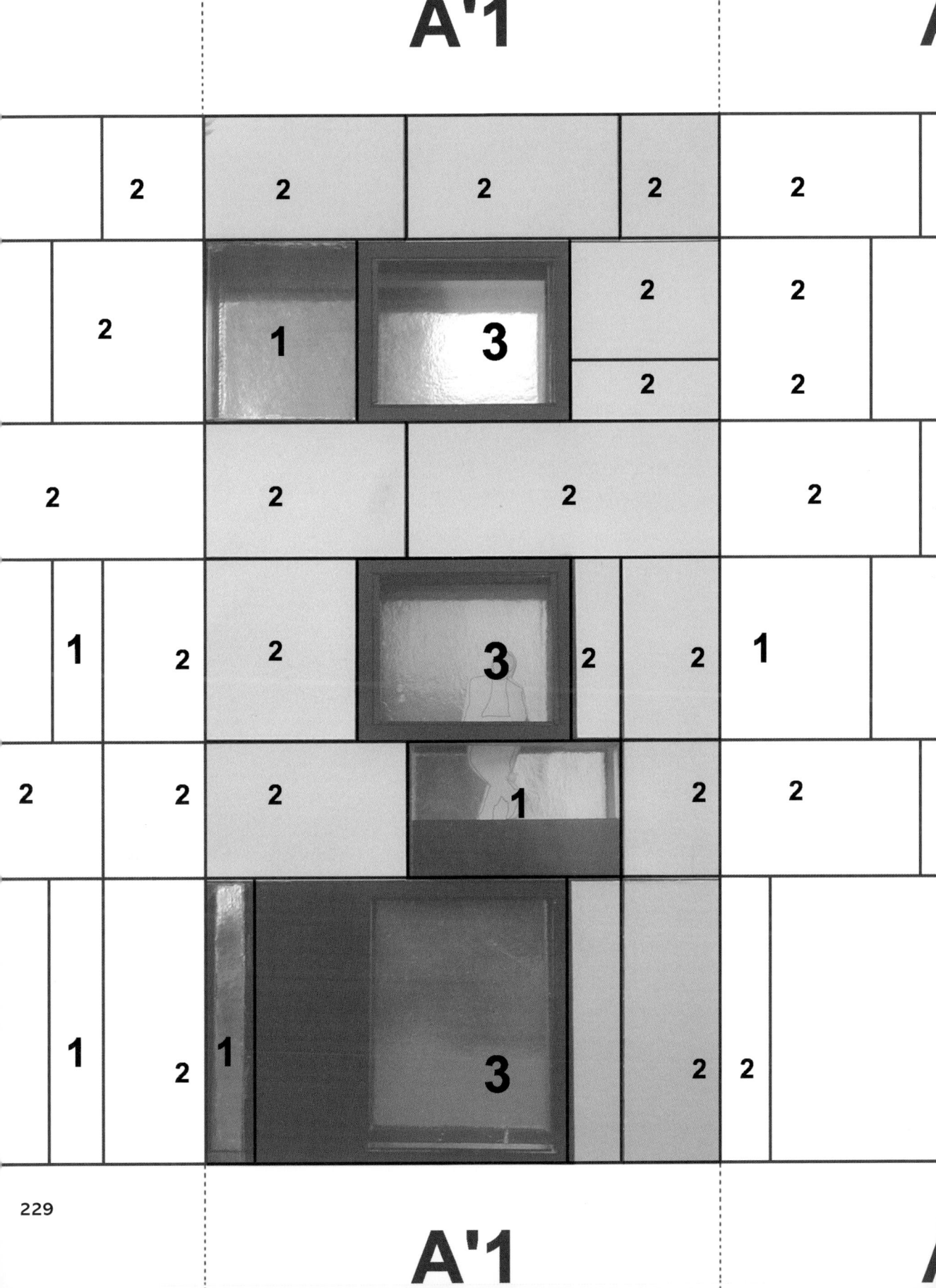

229

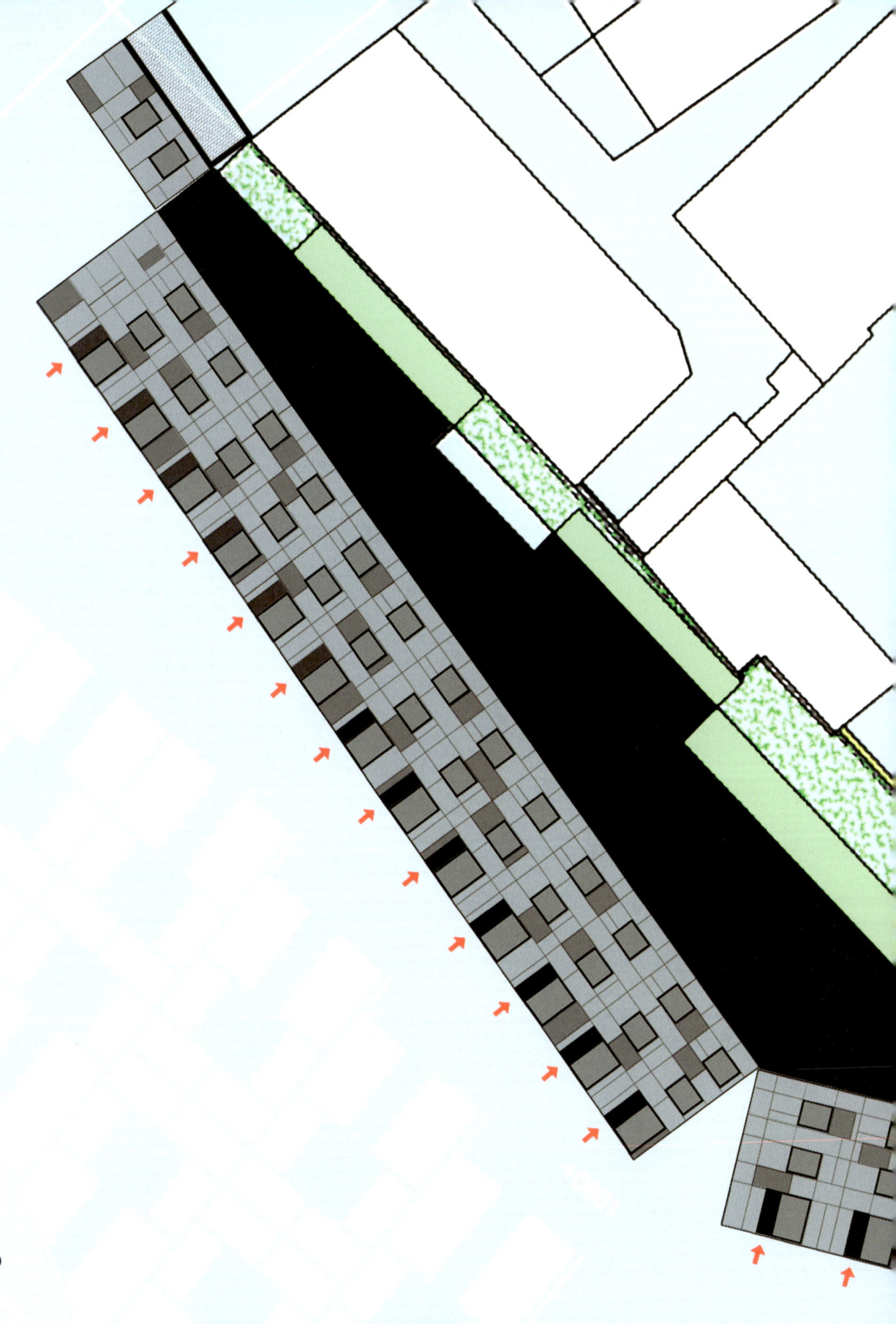

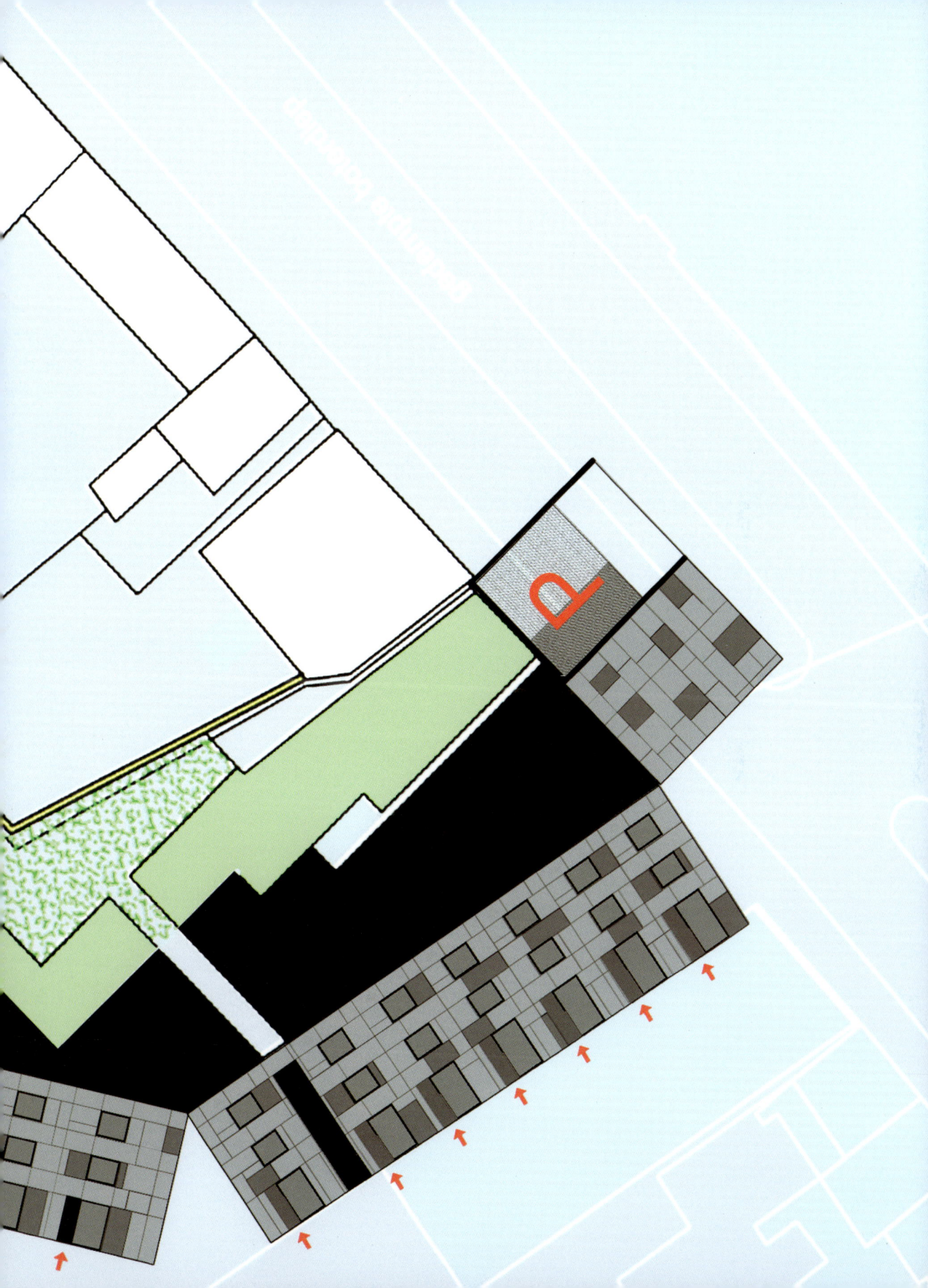

Suytkade
Seriality

The facades of this residential building in a new neighbourhood in Helmond designed by Soeters Van Eldonk Ponec are clad with gold–coloured aluminium shingles. Per storey they form a repeating pattern of five bands in four discrete heights. The windows also form a series, in which a tall window, two horizontally set oblong windows and four square windows of various size appear in a range of configurations. In addition, the facades are interrupted by open loggias and balconies measuring 3 x 3 metres, 6 x 3 and 3 x 6 metres, and an L–shaped type of 6 x 6 metres.

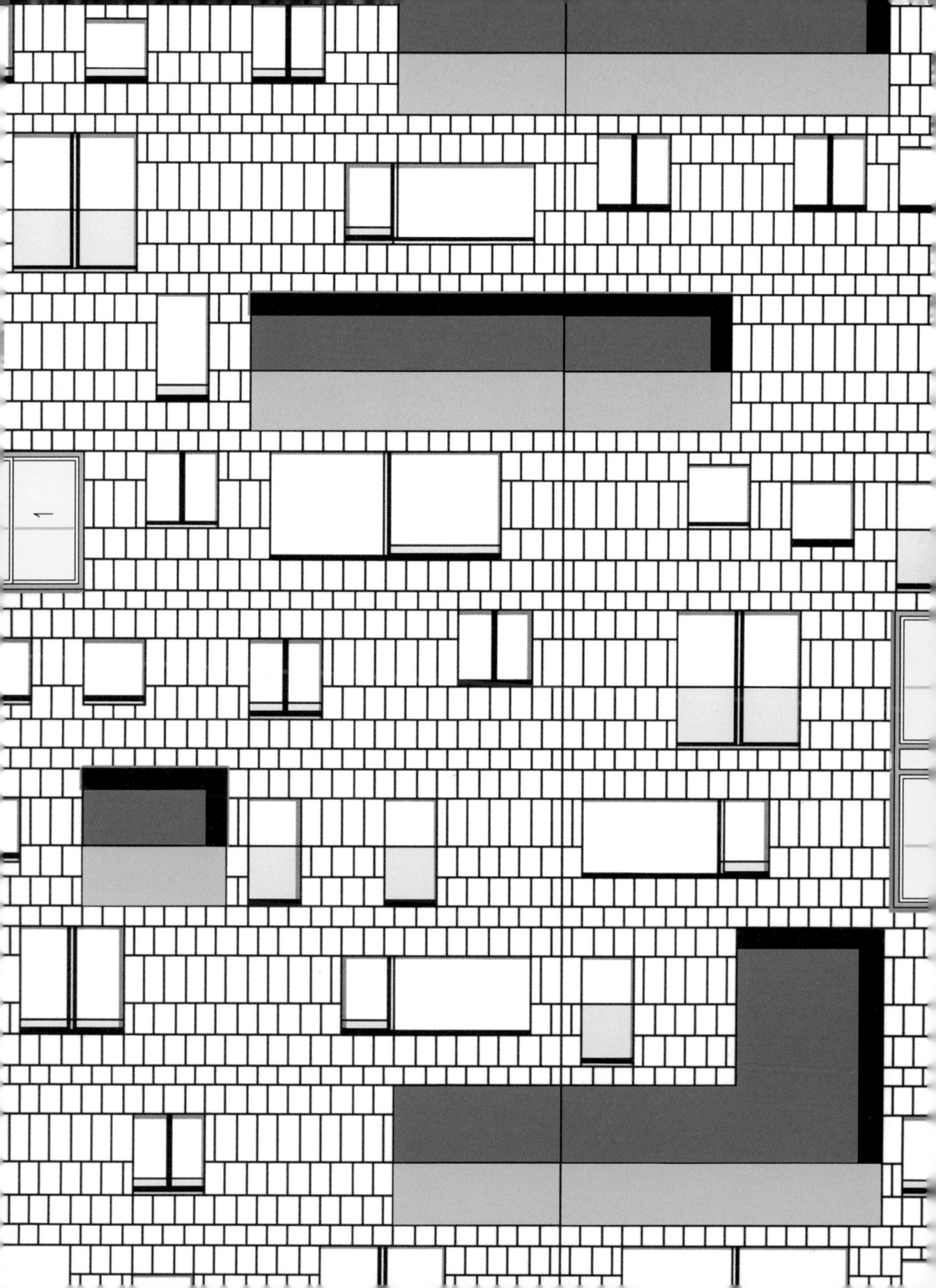

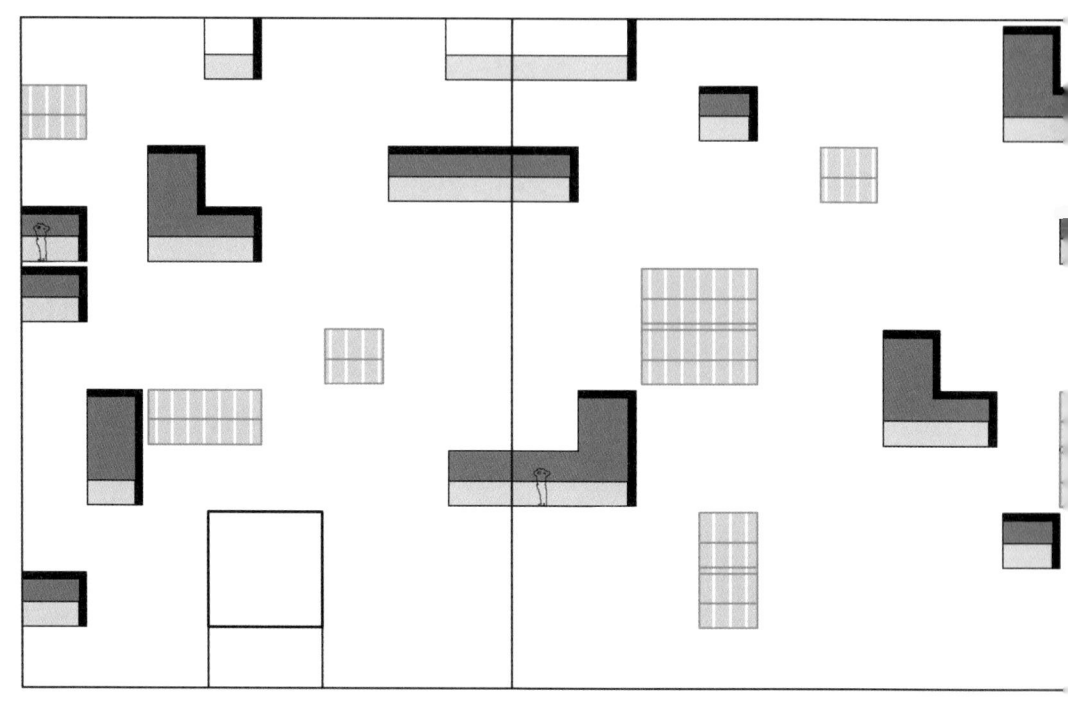

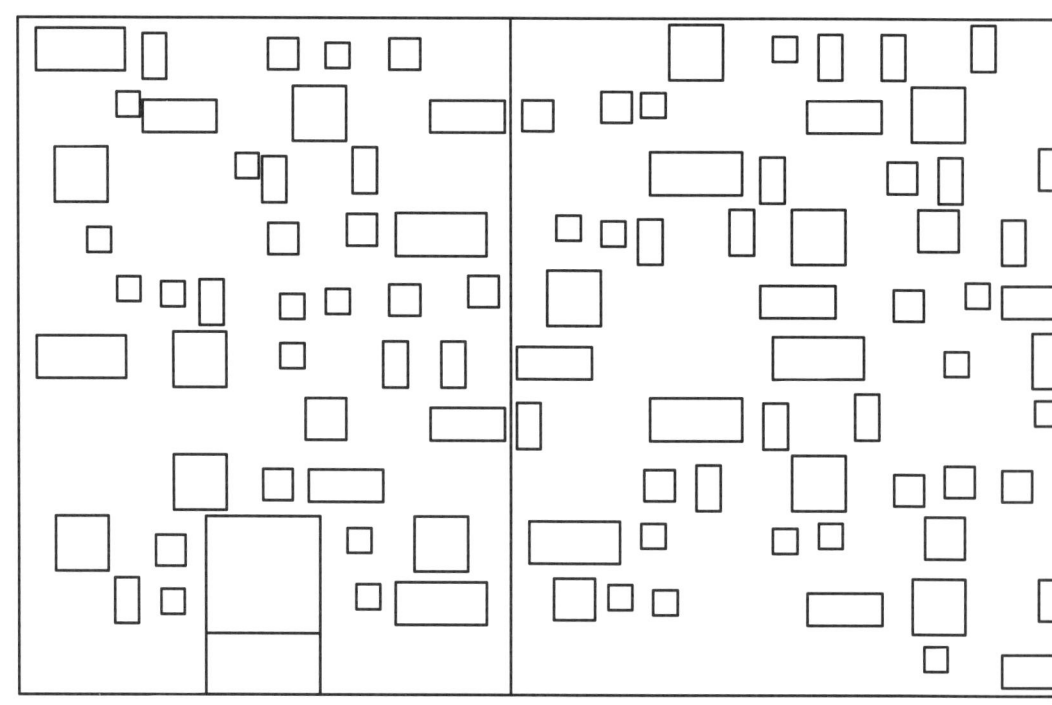

Witbrant
Seriality

In the patio houses within the urban plan designed by Jacq. de Brouwer for the Witbrant–Oost neighbourhood in Tilburg, the greenery, the living zones and the sleeping zones form a range of series which are recapitulated in various combinations and configurations. These series do not simply have consequences for the interior but also for the exterior: they establish rhythms in the extended length of the blocks.

sleeping zones

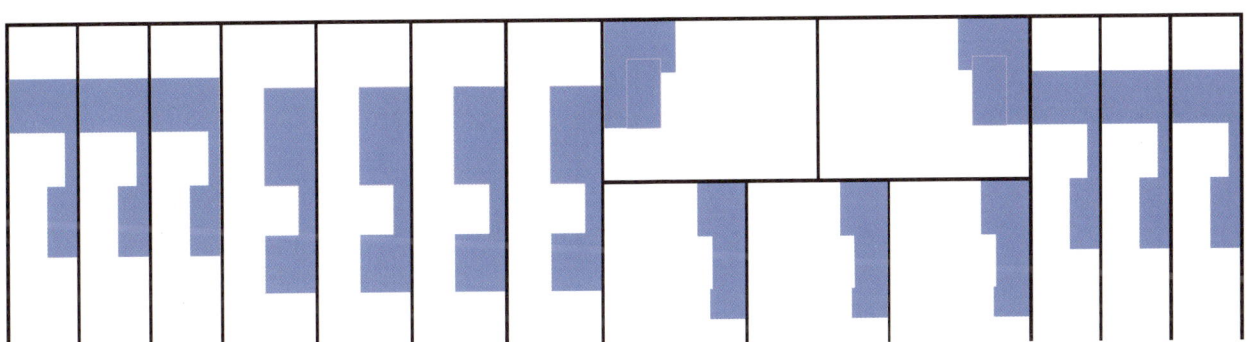

living zones

outside spaces

237

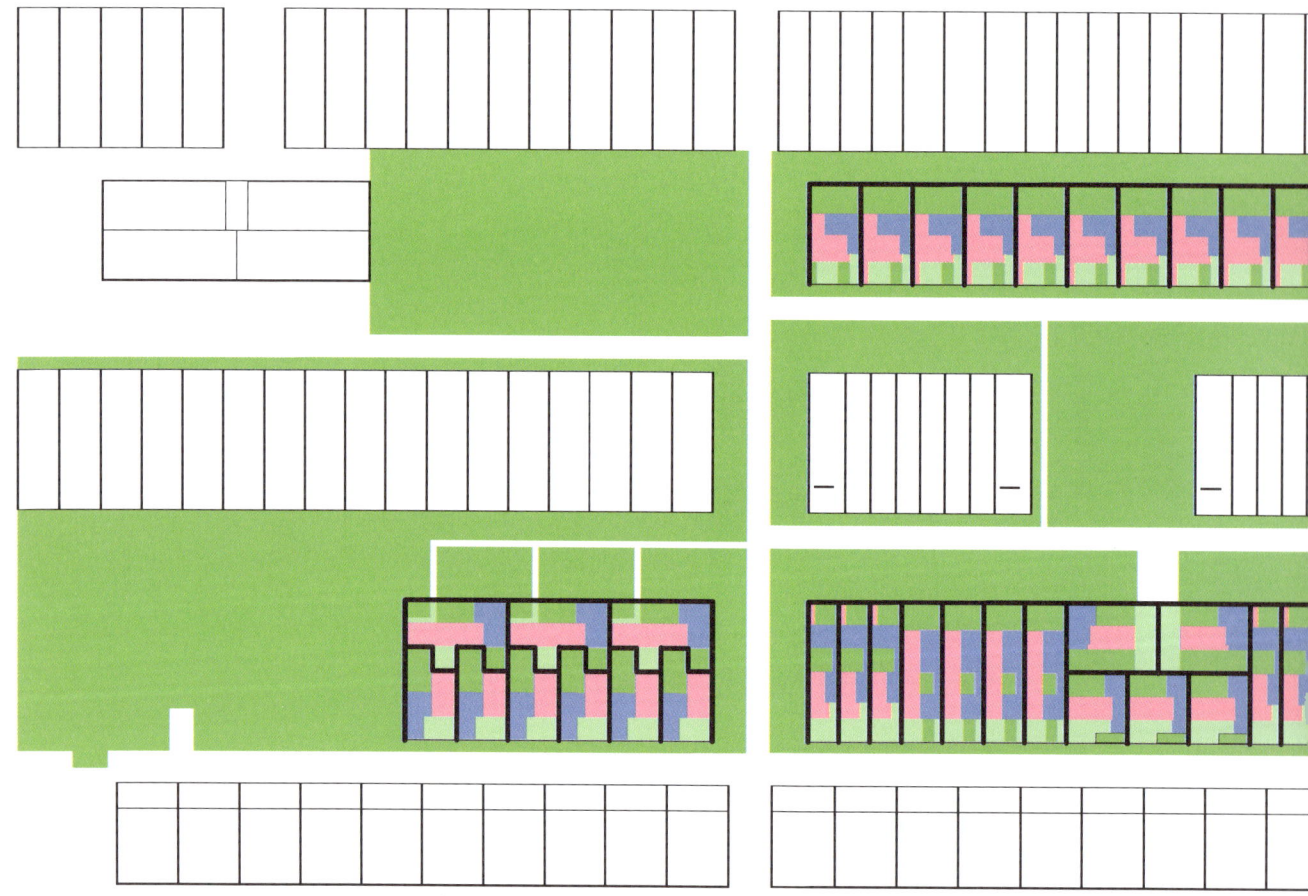

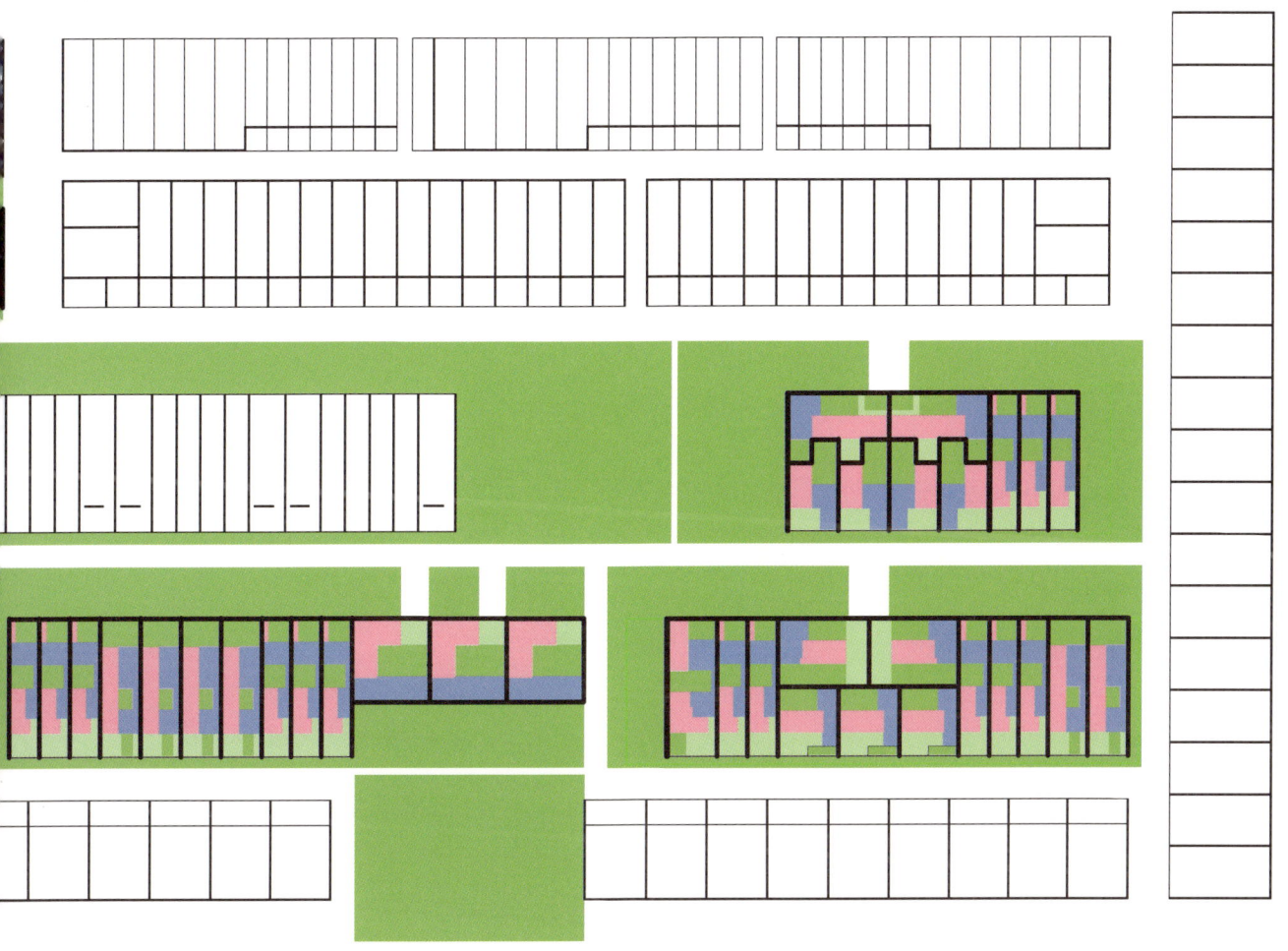

239

Haarlem
Seriality

The elevations and floor plans of the two extended housing blocks in Haarlem's Zuiderpolder development are composed of uniform layers which each follow their own rhythm. The slender columns, the stairways and the configuration of large and small openings are all repeated in the front elevation, but each at a different interval. The floor plans of the dwellings are oriented in a single direction as a succession of zones, proceeding from public to private. That is coupled with a progression from closed (on the street) to open spaces (on the garden side), with a long corridor in between. In the other direction the floor plan consists of a systematic succession of rooms. The garden elevation coordinates with this system in an autonomous manner.

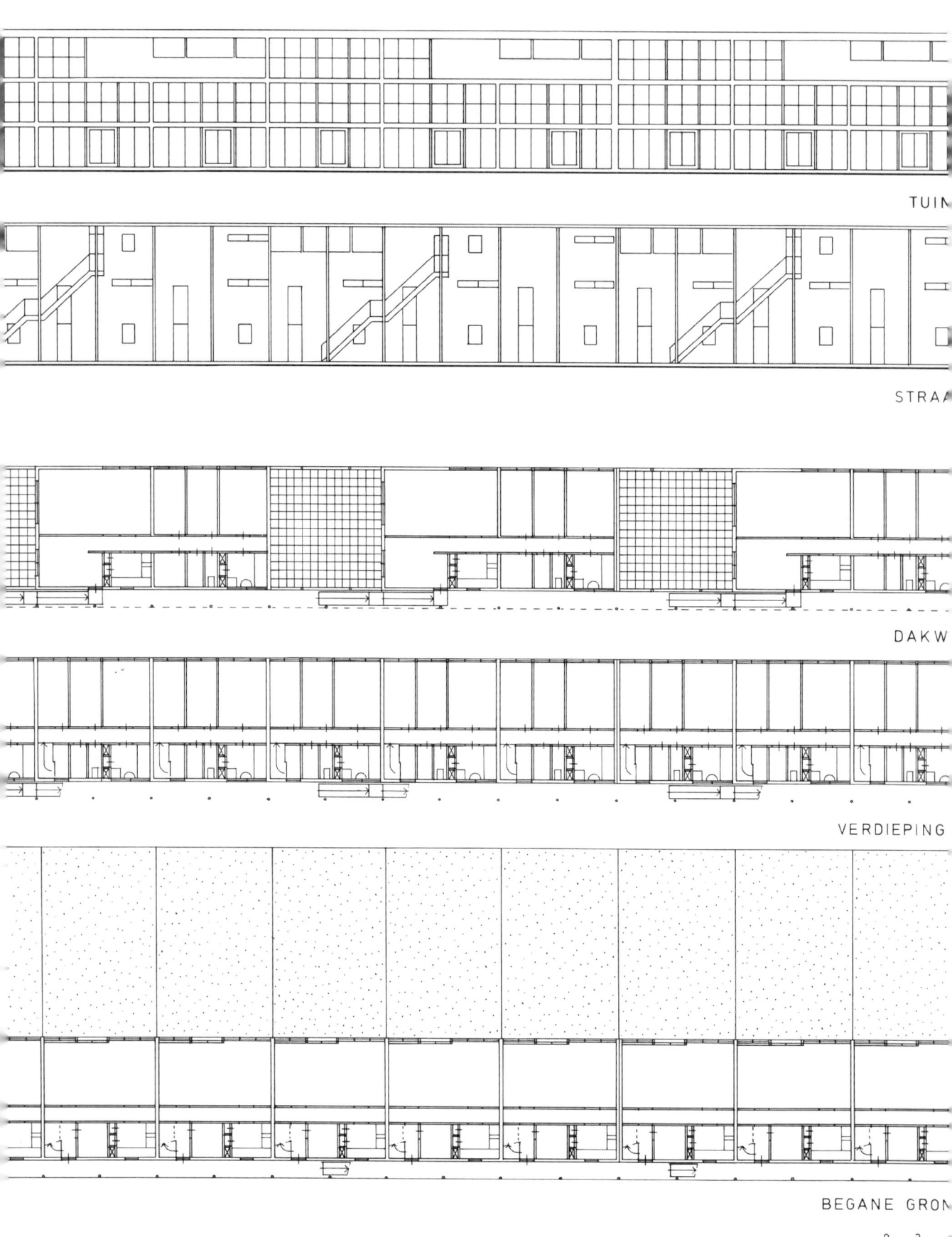

TUIN

STRA

DAKW

VERDIEPING

BEGANE GROM

0   2

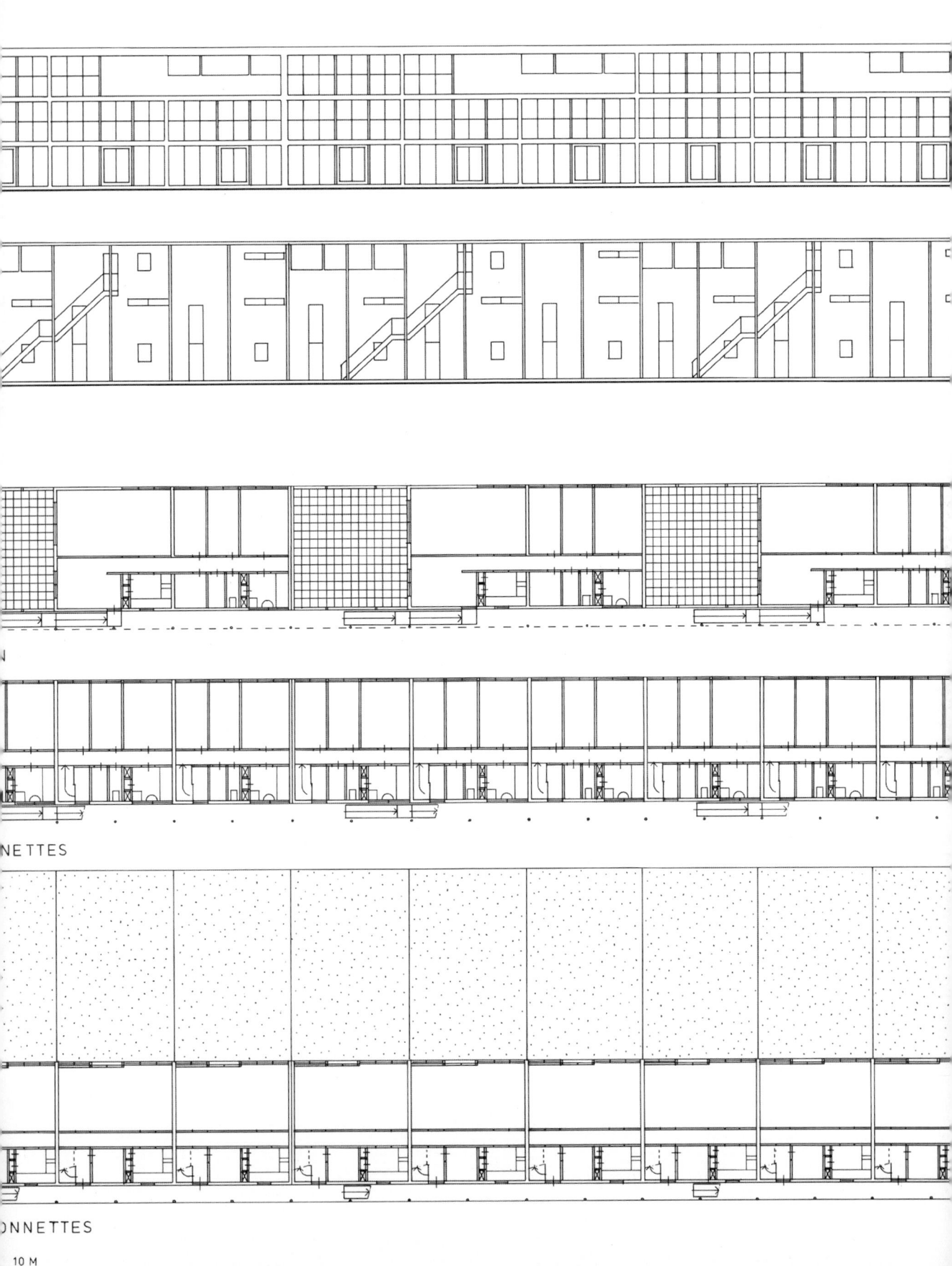

NETTES

NNETTES

10 M

Het Funen
Seriality

The power of the residential block at Het Funen in Amsterdam is the ten–fold repetition of a unique figure. The building consists of ten similar dwellings arranged in complementary pairs and interlocked by staggering them in width and height. Together the ten dwellings form a rectangular block surrounded by a homogeneous facade.

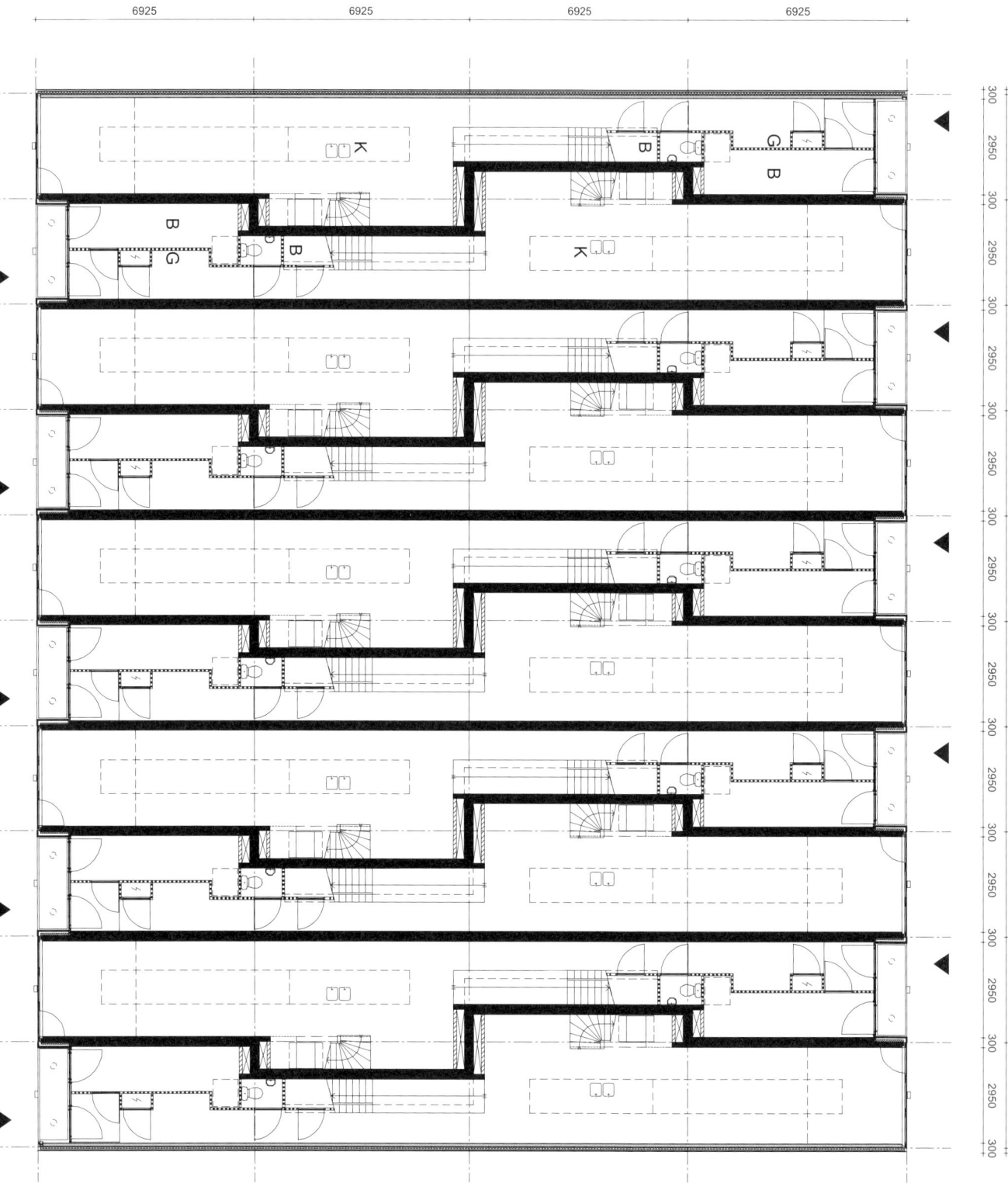

248

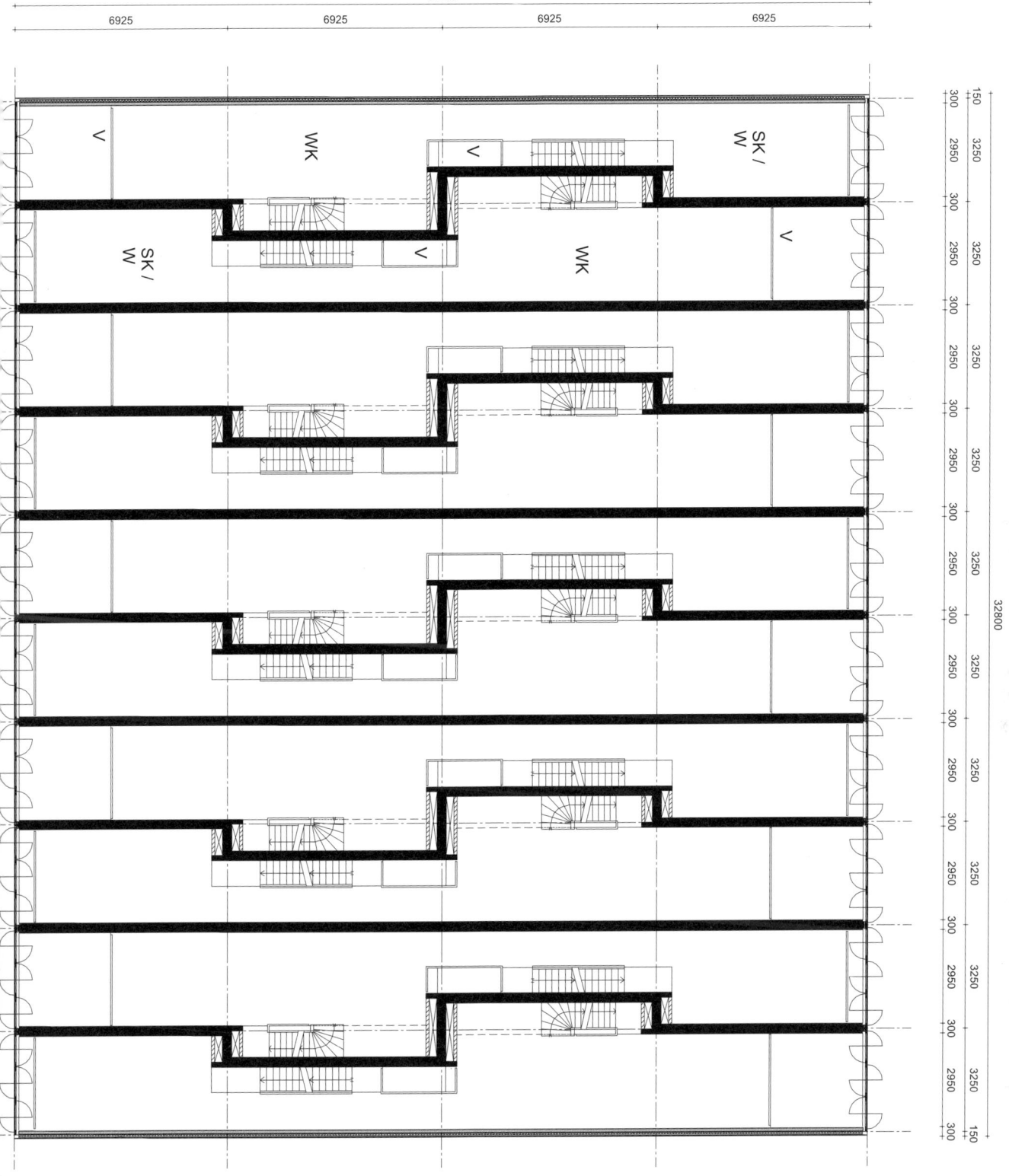

27700

6925 6925 6925 6925

V

WK

V

SK / W

SK / W

V

V

WK

V

32800

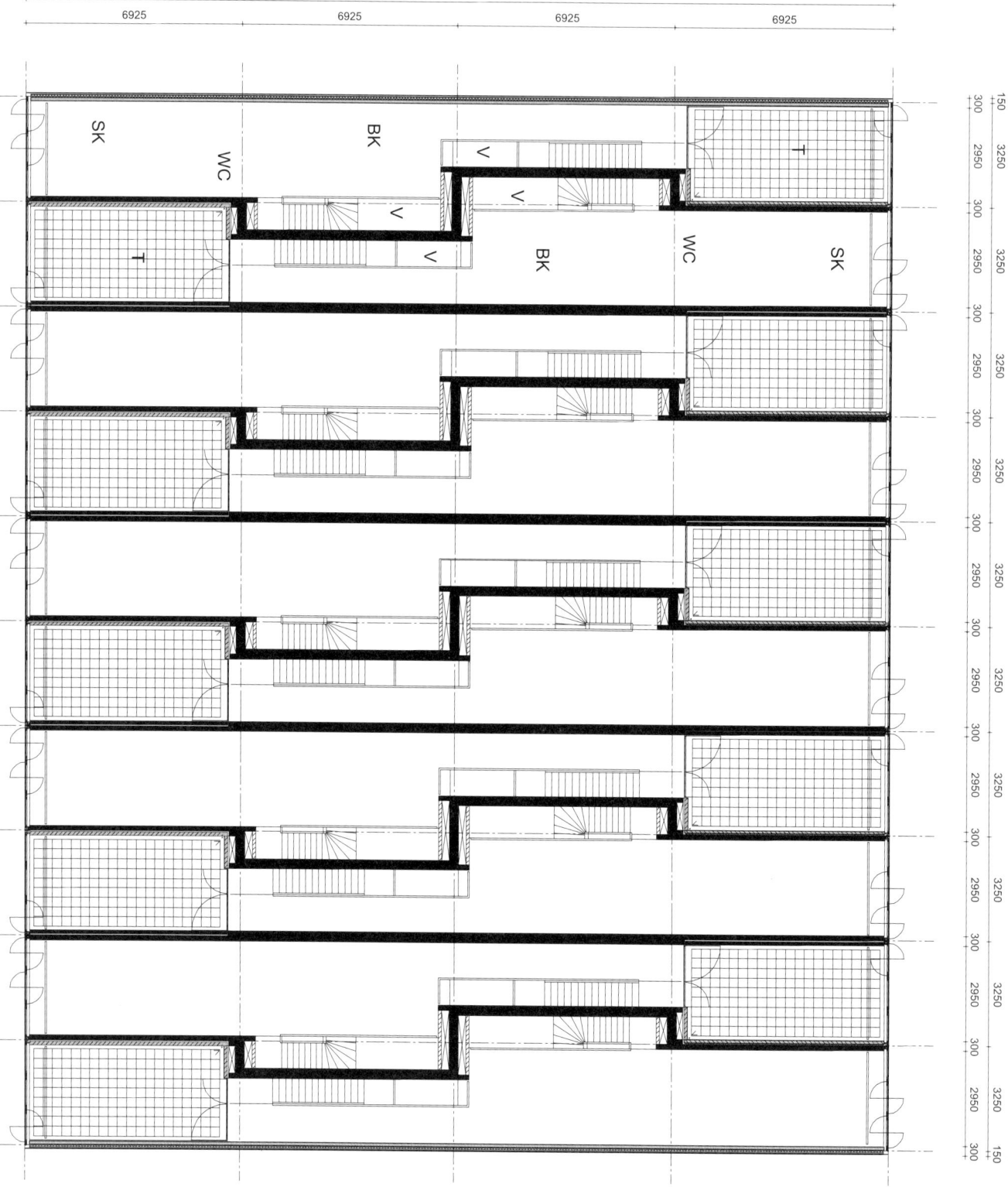

250

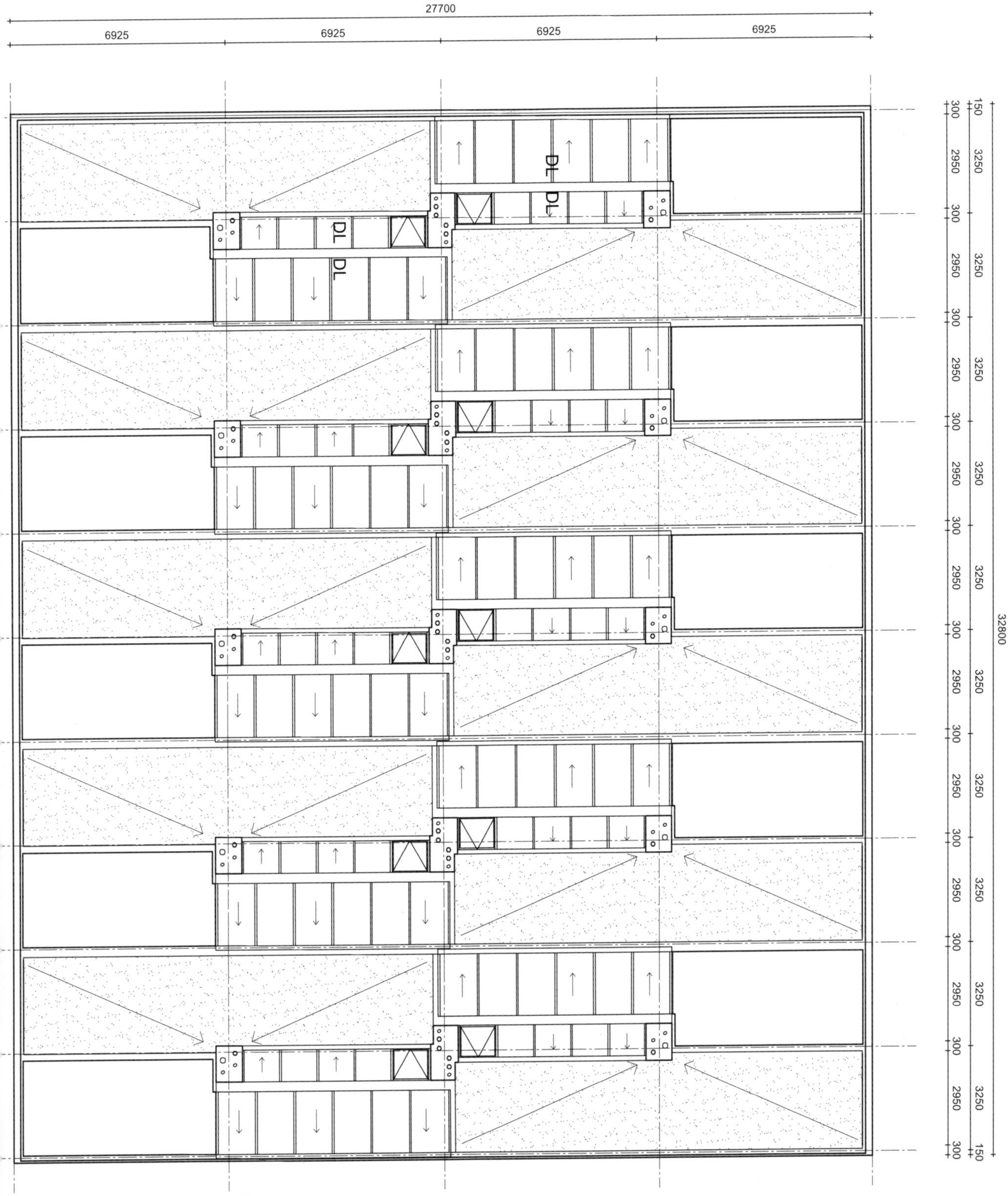

27700

6925 6925 6925 6925

DL
DL
DL
DL

32800

150
300
3250
2950
300
3250
2950
300
3250
2950
300
3250
2950
300
3250
2950
300
3250
2950
300
3250
2950
300
3250
2950
300
150

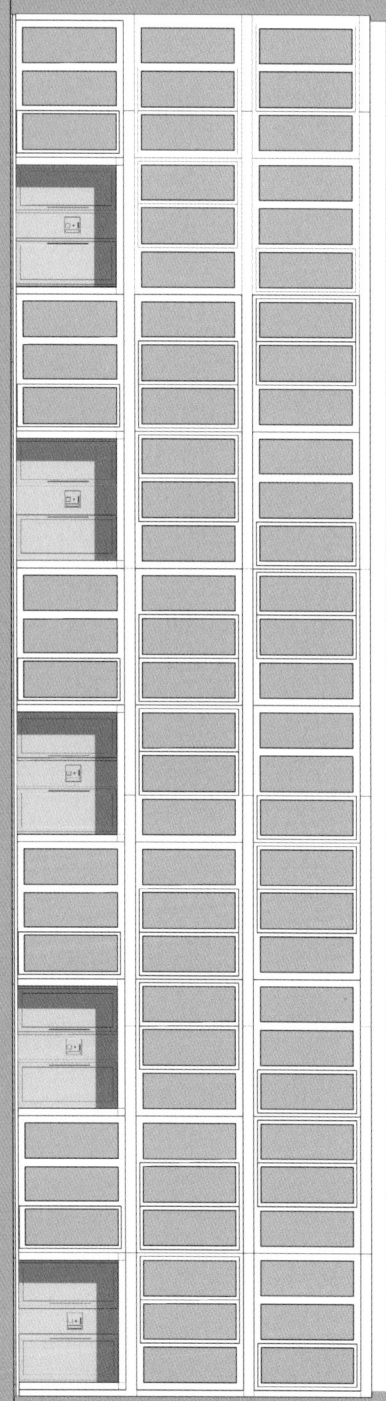

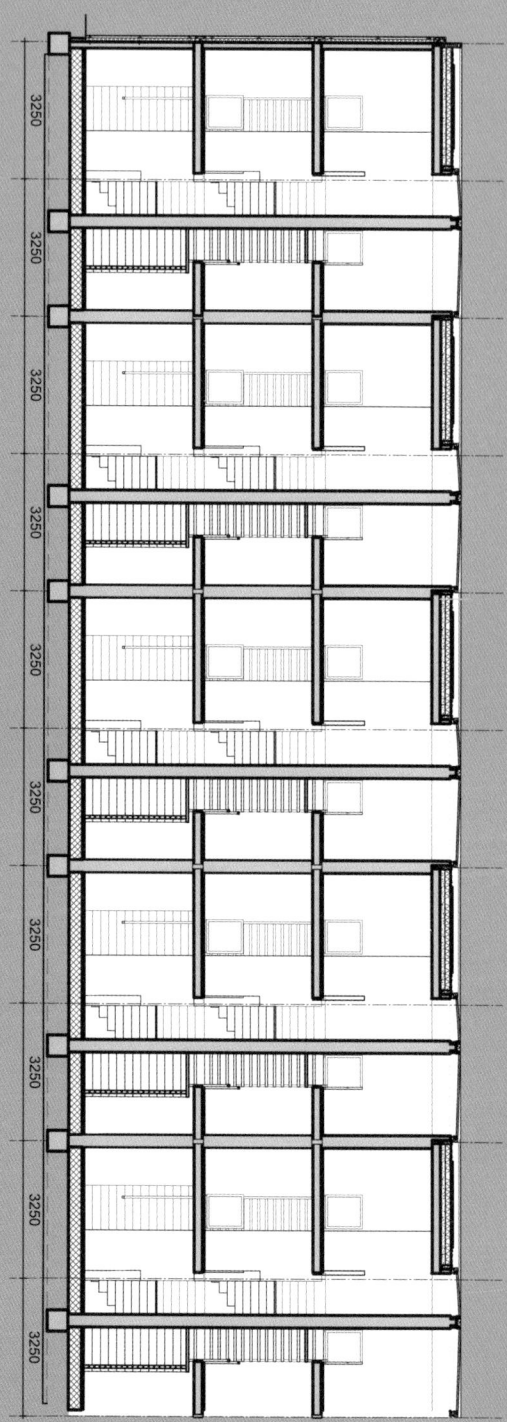

Almere 6A
Seriality

The wide facades of Block 6A, a residential building in the centre of Almere Stad, are covered with vertically hinged panels, which the inhabitants can fully or partially open at will. On the elevation facing the cinema in Block 6, the facade can be completely closed off. On the sunny side, the only sections without shutters are those in front of the glazed walls. The shutters make it possible to isolate the public nightlife and the private lives within the dwellings. The shutters result in variable patterns of open and closed sections, transparency and non–transparency.

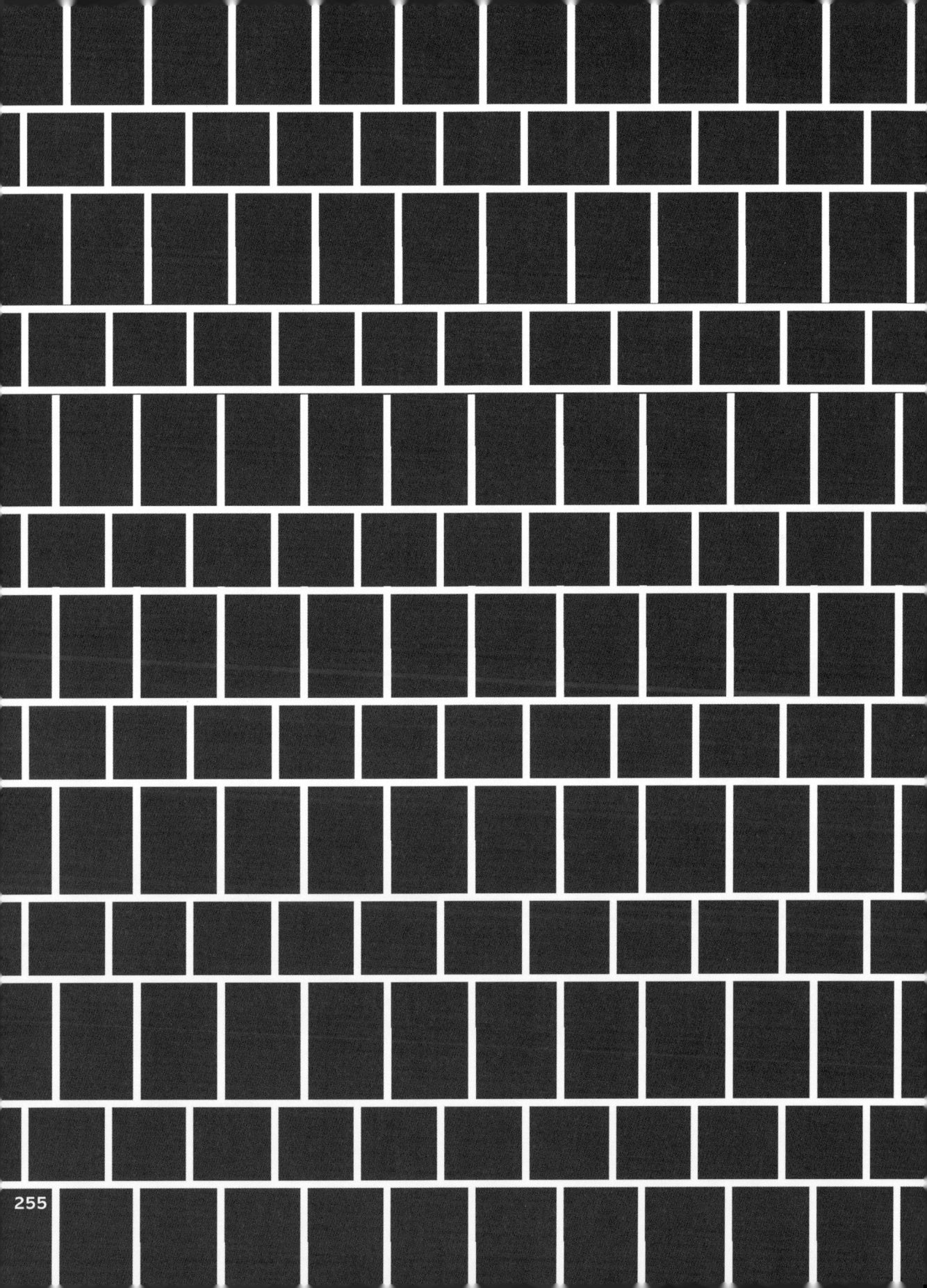

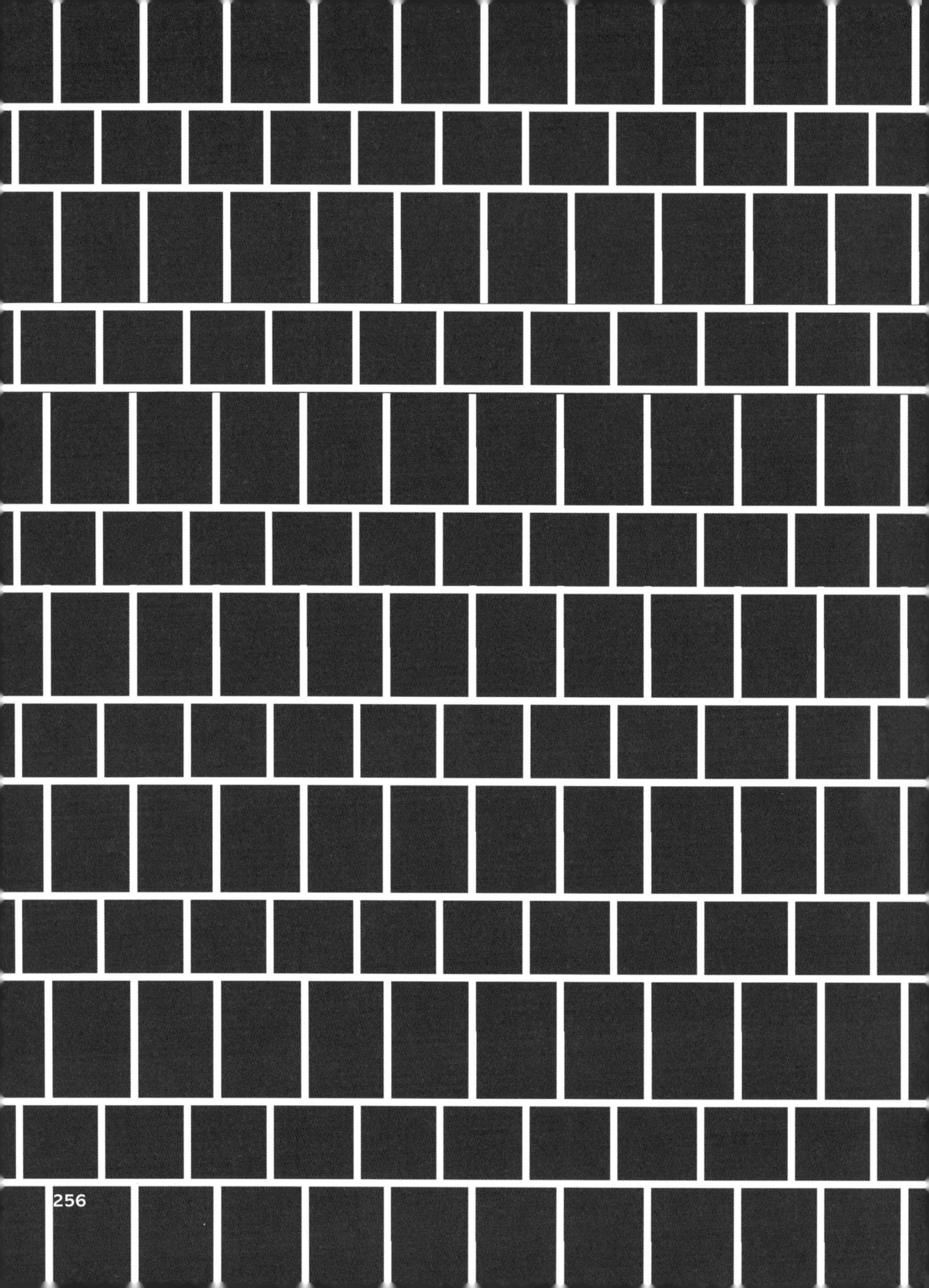

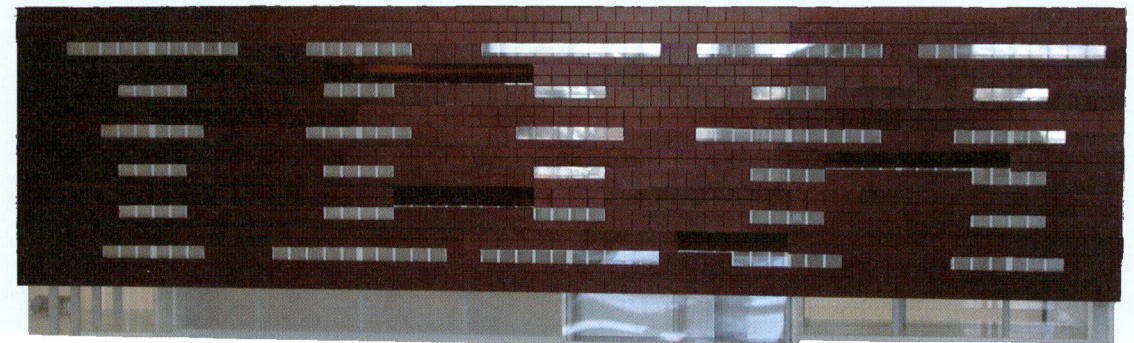

Stramanweg
Seriality

The colour pattern of the brickwork surfaces varies for each of the three blocks in a series of five on a trunk road in the Amsterdam borough of Zuidoost. The buildings are therefore not identical, but form a series of related objects. The buildings do not have front, side or rear elevations, but an all–round envelope that varies for each block and facade. The elevations are composed of three overlapping patterns: verandahs which are also the rear entrances to the dwellings, surfaces of shiny black brickwork and storey–high glazed walls, which do not coincide with dwellings but continue along in front of them. This means that each dwelling has a different incidence of light and a specific character. Each block also has a vertical segmentation, as the storeys are edged with a horizontal steel band.

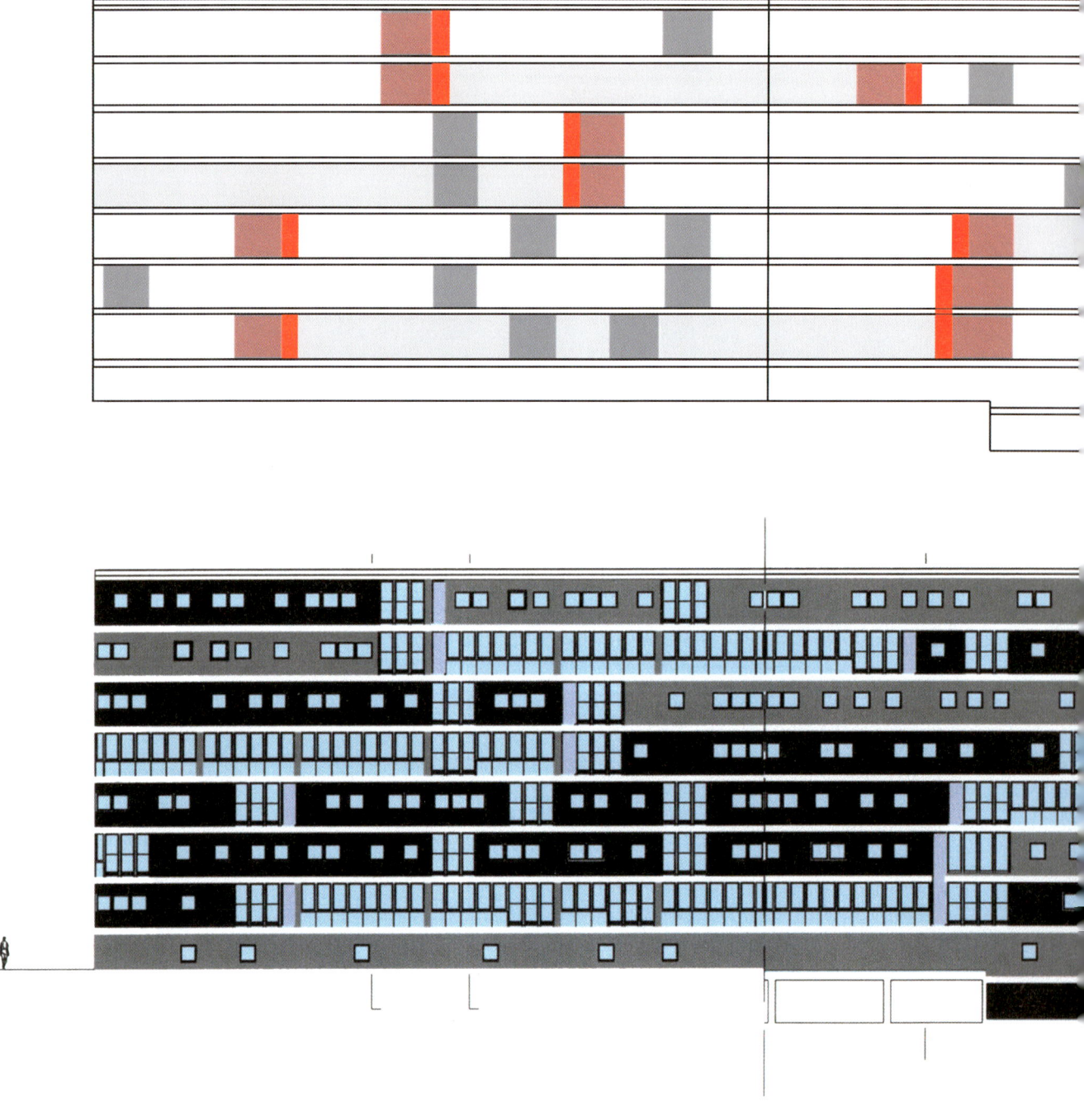

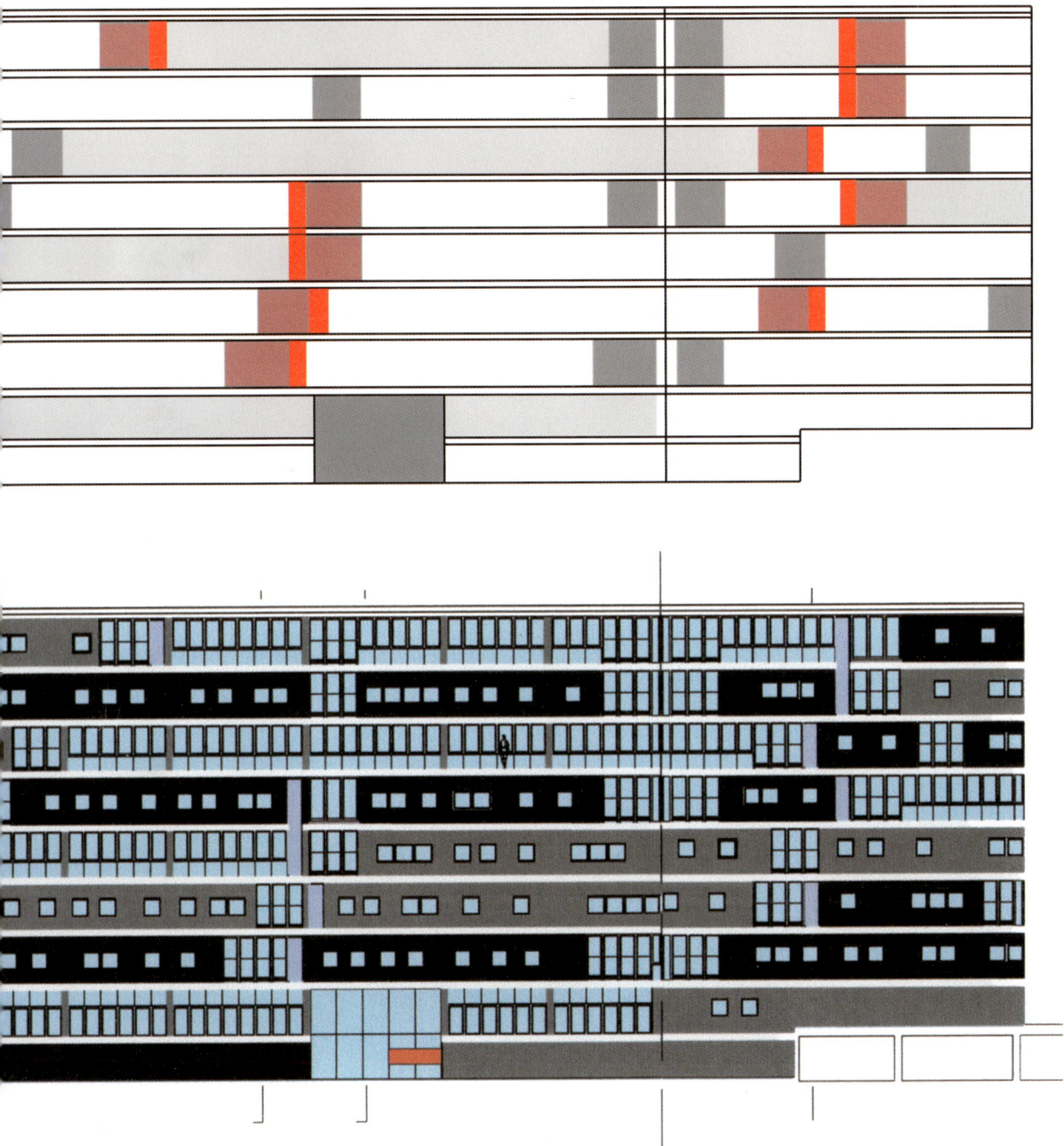

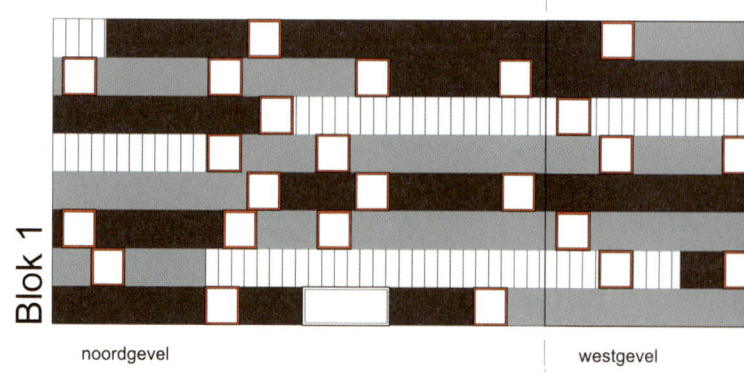

Blok 1

noordgevel   westgevel

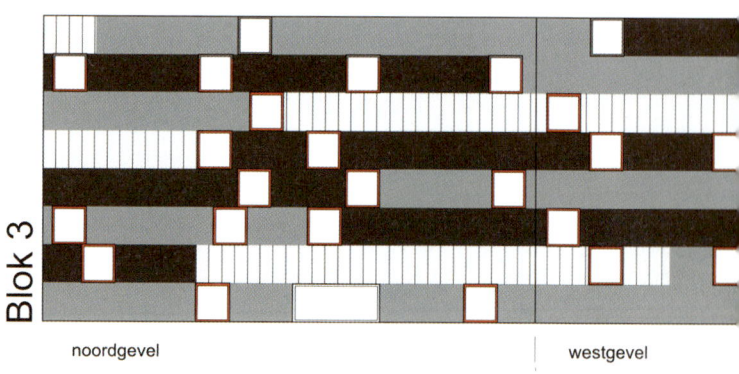

Blok 3

noordgevel   westgevel

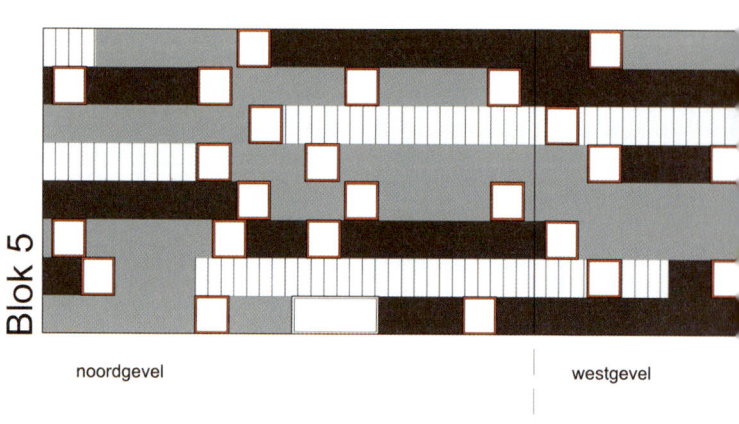

Blok 5

noordgevel   westgevel

266

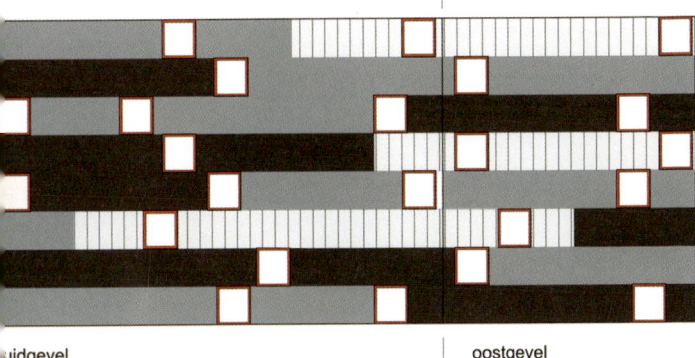

uidgevel   oostgevel

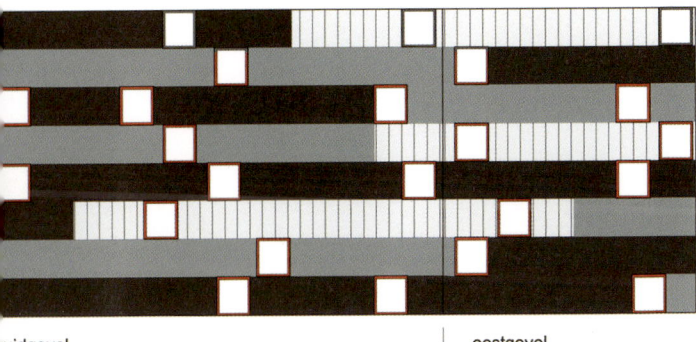

uidgevel   oostgevel

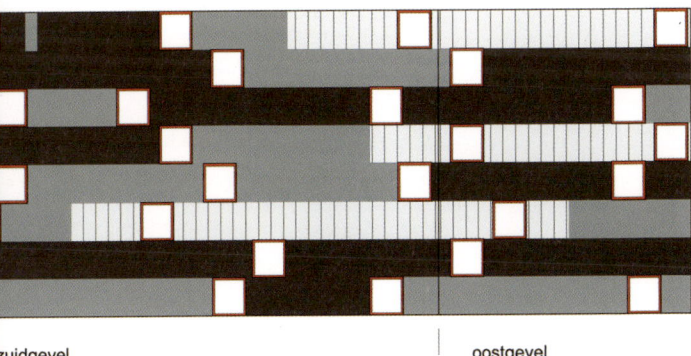

zuidgevel   oostgevel

267

Chassé 100 x 100
Seriality

The 100 x 100 building in the Chassé Park in Breda contains a single type of dwelling, composed of the same set of elements. What varies is the position of the kitchen on the entrance level (one metre above street level), of the stairs to the cellar and the upper floor, of the bathrooms (one or two per unit), and the rooftop terraces (also one or two per unit) on the upper floor. These elements are organized in various series, each with its own pattern, which together form the block. The facades of timber and stone on the upper floors establish another series, and the facades of each dwelling are variously subdivided by the window openings, open and closed sections. Again there is a wooden sliding wall to the front of the facade, so a range of changing figurations can arise within the series.

25m

0

269

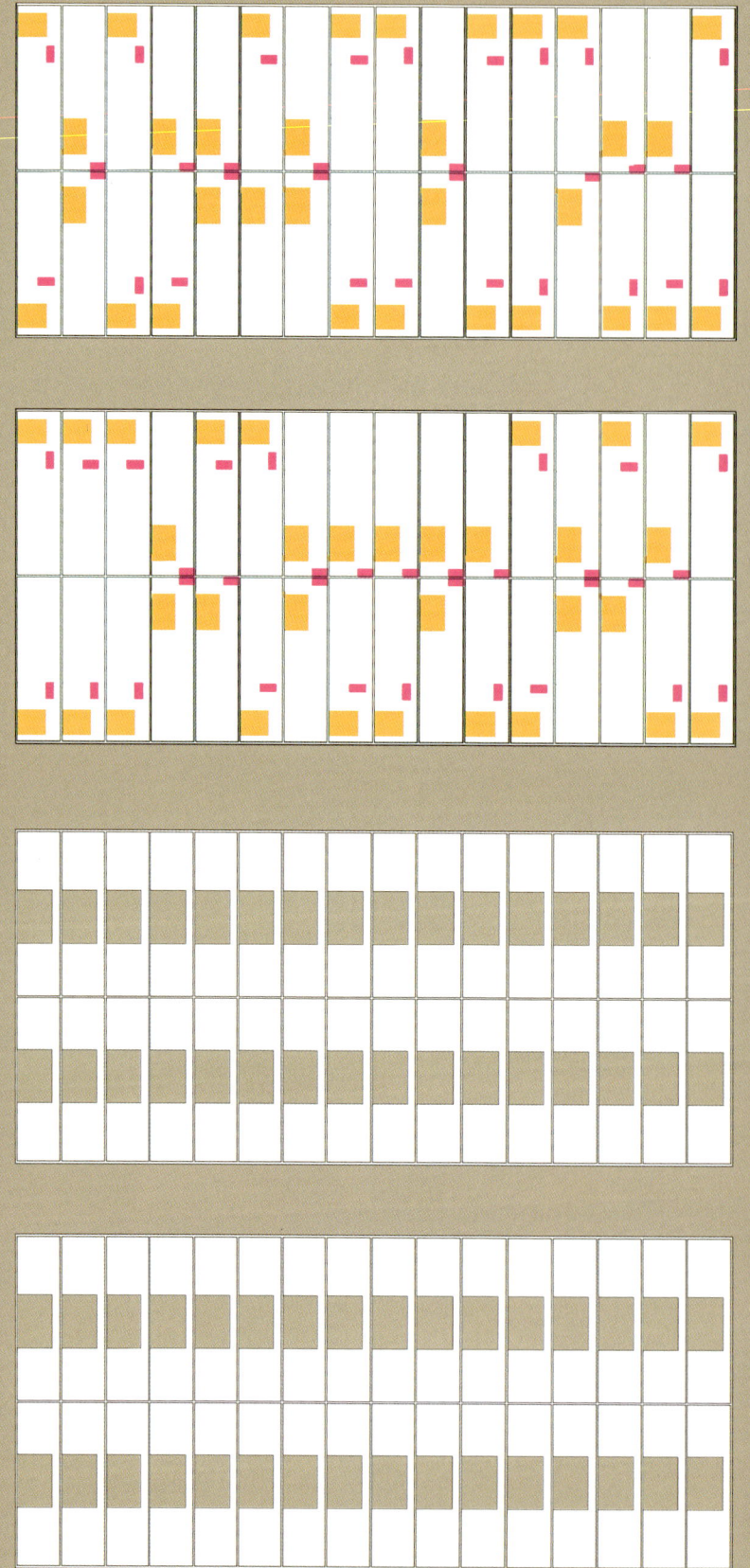

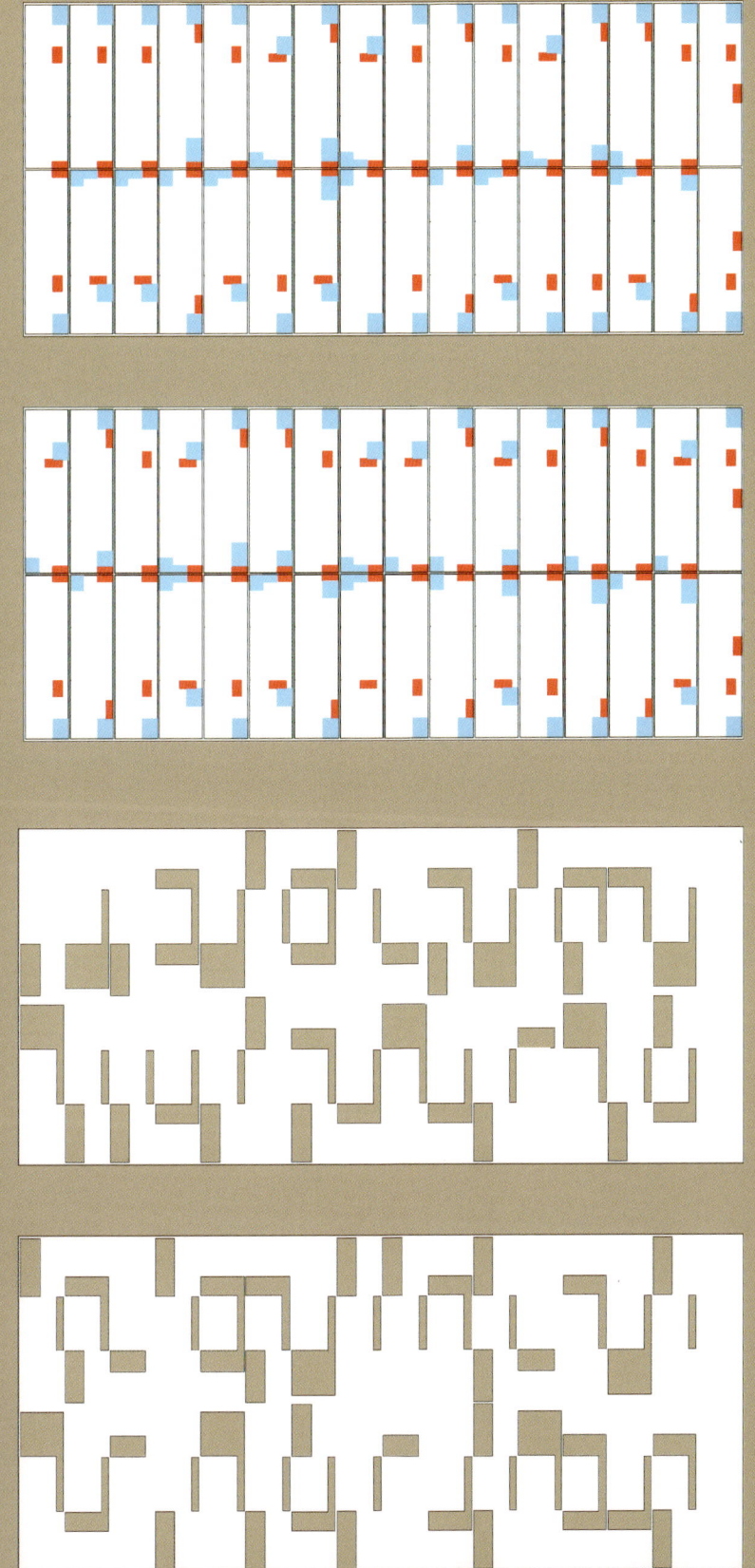

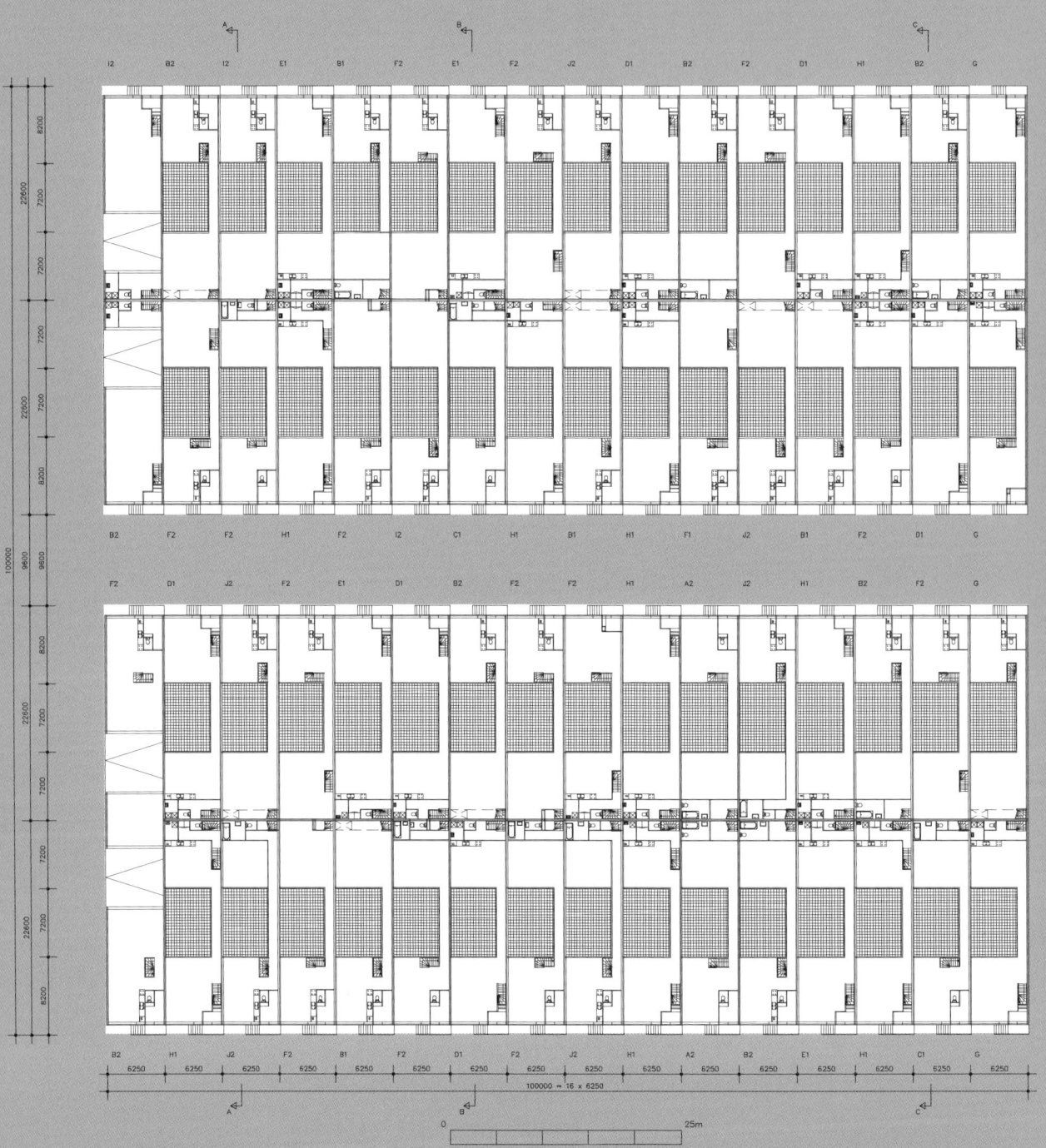

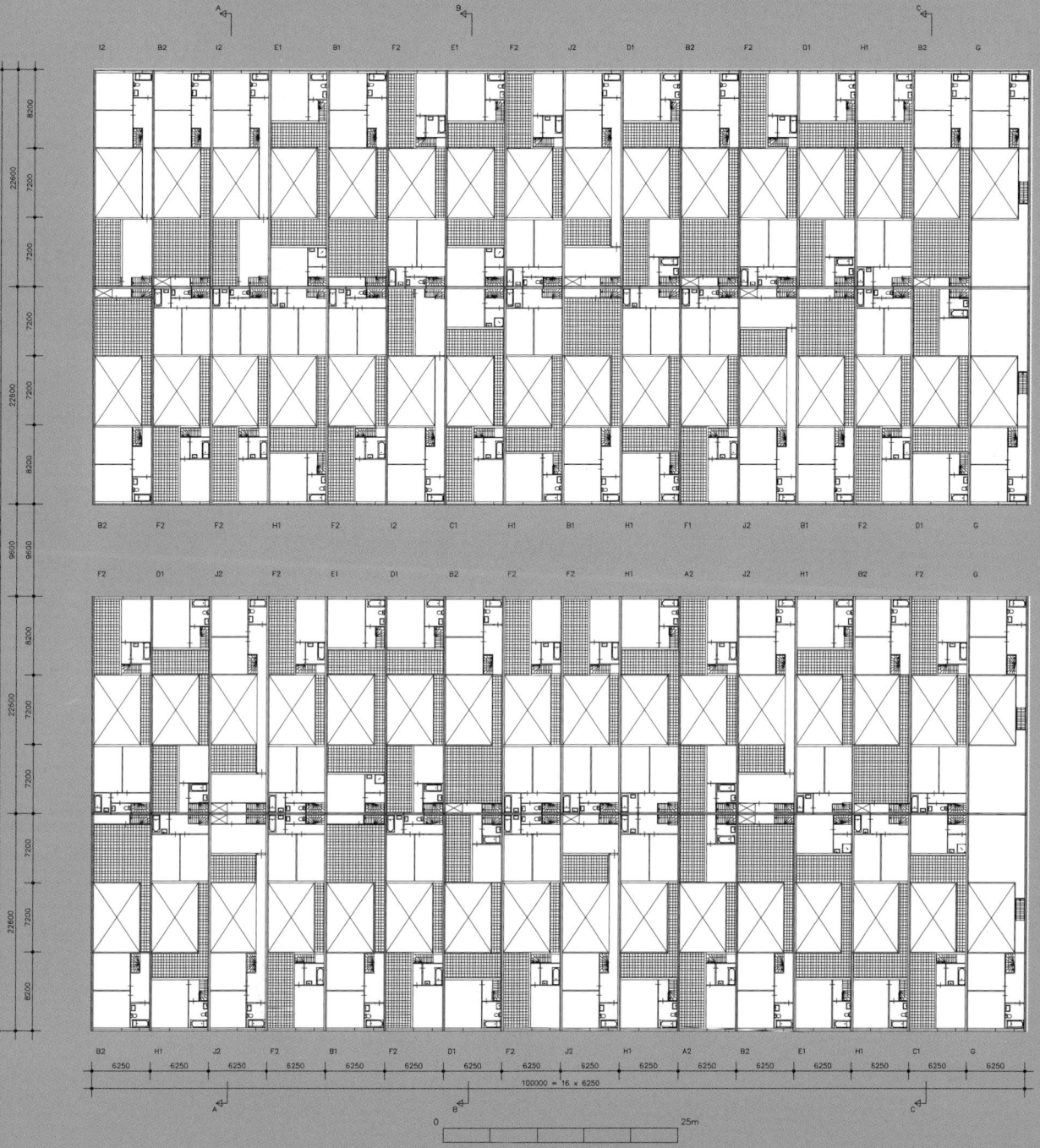

Muiderstraat
Seriality

A pattern of elements with louvres was added to the continuous glazed facade of the office building on the corner of the Muiderstraat and Rapenburgerstraat in Amsterdam. This has a unifying effect and results in a varying incidence of light in the office spaces.

De Aker
Seriality

The housing project in De Aker in Amsterdam is organized as serial variations on a single–storey rectangular field. Various patterns, each with a specific orientation and rhythm, are superimposed on that field resulting in a distinctive spatial effect. The field is subdivided into blocks. Constricting the blocks has resulted in inner streets of varying widths. Setting the patios of all the houses on the south side of the plot and adding a second storey along the inner street has resulted in patio houses on the south side of each block that are of a completely different type to those on the north side. This asymmetry also means that the street elevations standing opposite each other are completely different. • The series of finishing details results in each dwelling within a block being different, with as a constant the balance between views and privacy, between looking out towards the outside world and the focus on the private world of one's own patio.

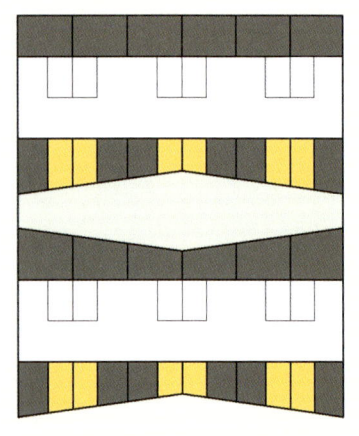
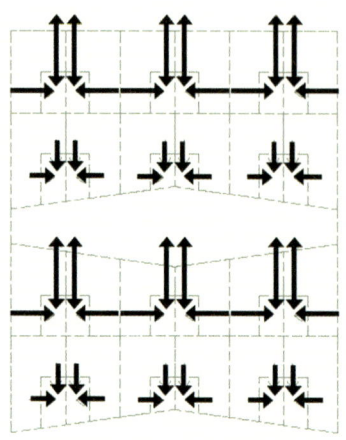
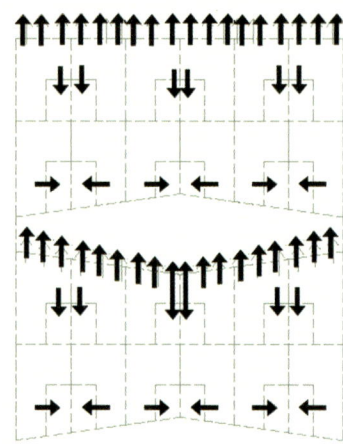

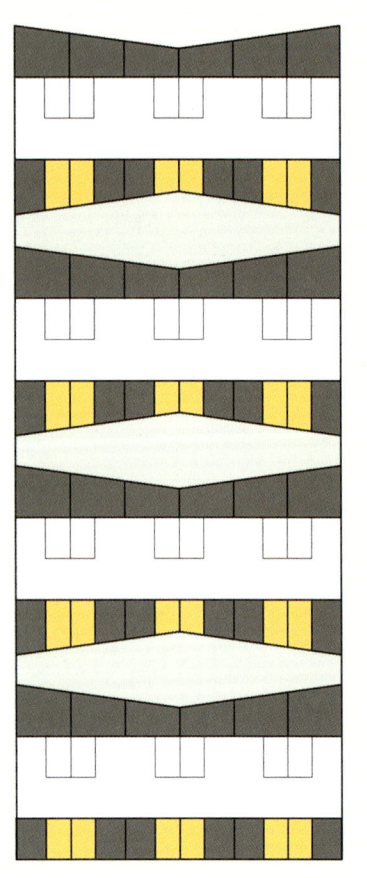
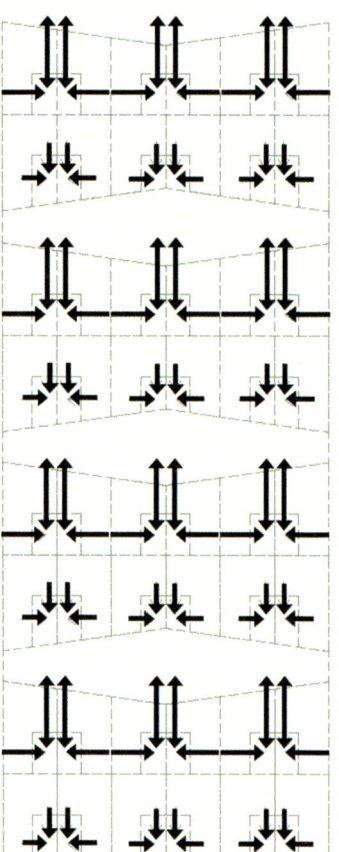
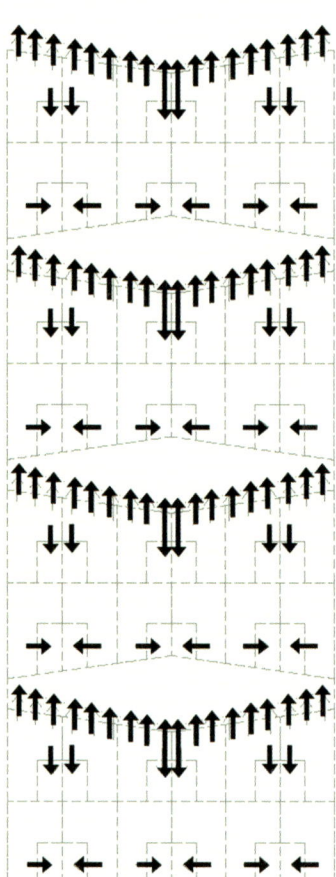

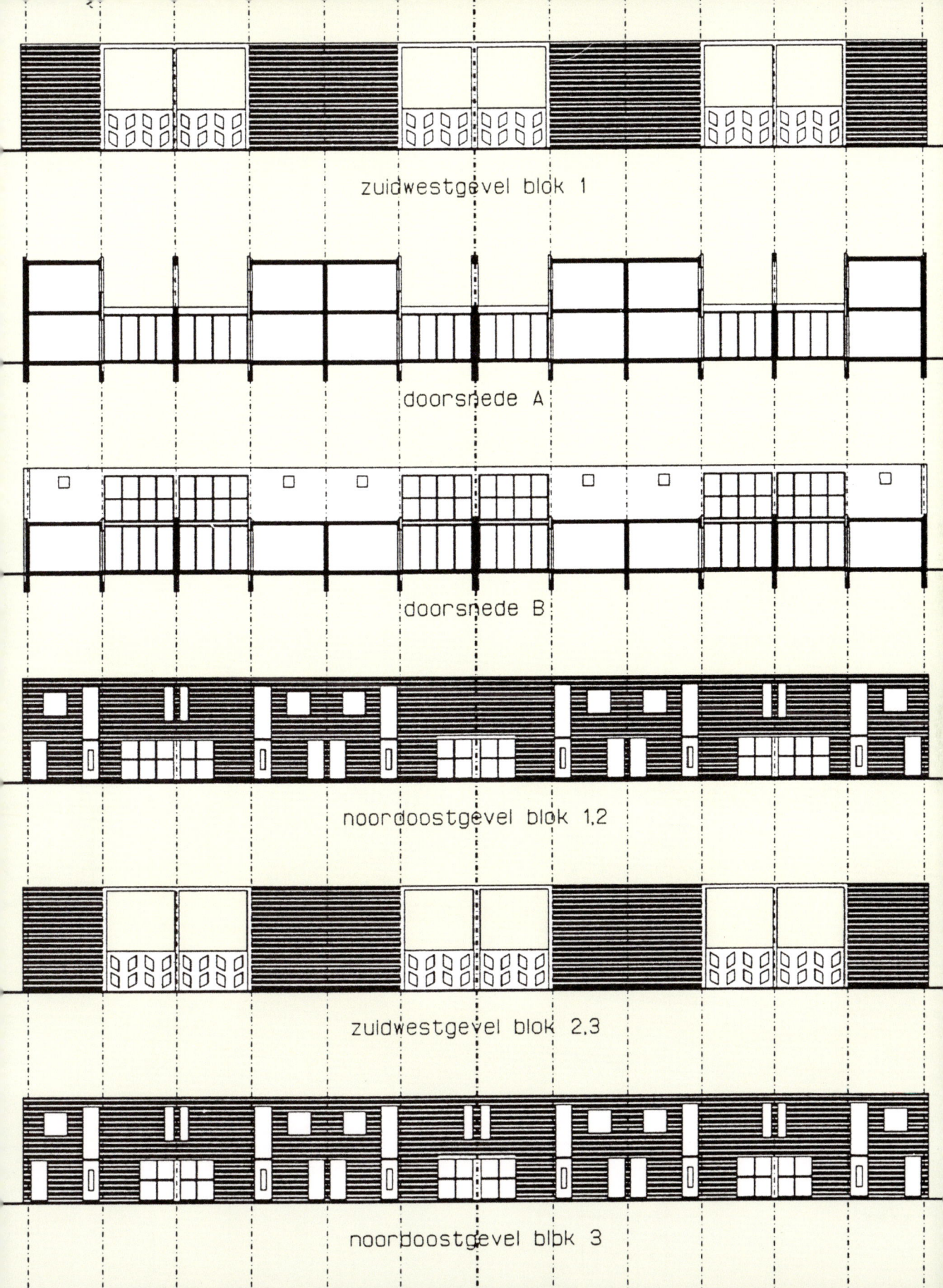

zuidwestgevel blok 1

doorsnede A

doorsnede B

noordoostgevel blok 1,2

zuidwestgevel blok 2,3

noordoostgevel blok 3

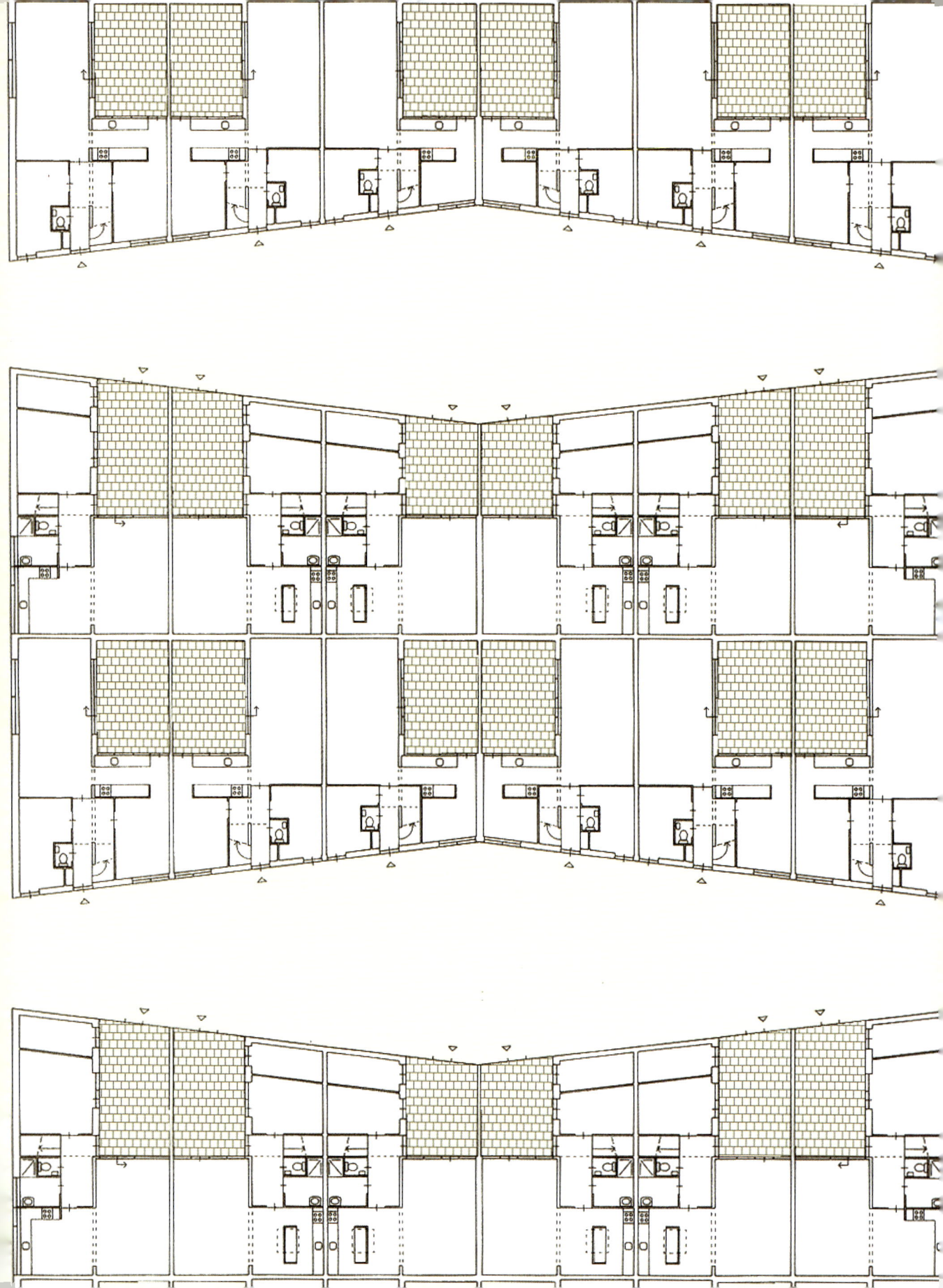

Erasmusgaarde
Seriality

In Erasmusgaarde in The Hague's Zuidwest redevelopment area, the blocks have been subdivided into a series of houses by giving them eye–catching doors and introducing clear–cut vertical expansions. The houses are characterized by a repetition of elements with a fixed subdivision of 4.20, 4.80 and 6.00 metres. The front door of all the dwellings is set at the right of the facade, with a canopy that houses street lighting as a constant element, continuing at one and the same height even if the front door is set half a storey higher. • The various spatial configurations of the dwellings can also be read as a series, each time with a progression from the more public functions at street level (such as the kitchen/diner) to the most private sections of the dwelling upstairs (the bedrooms and the terrace).

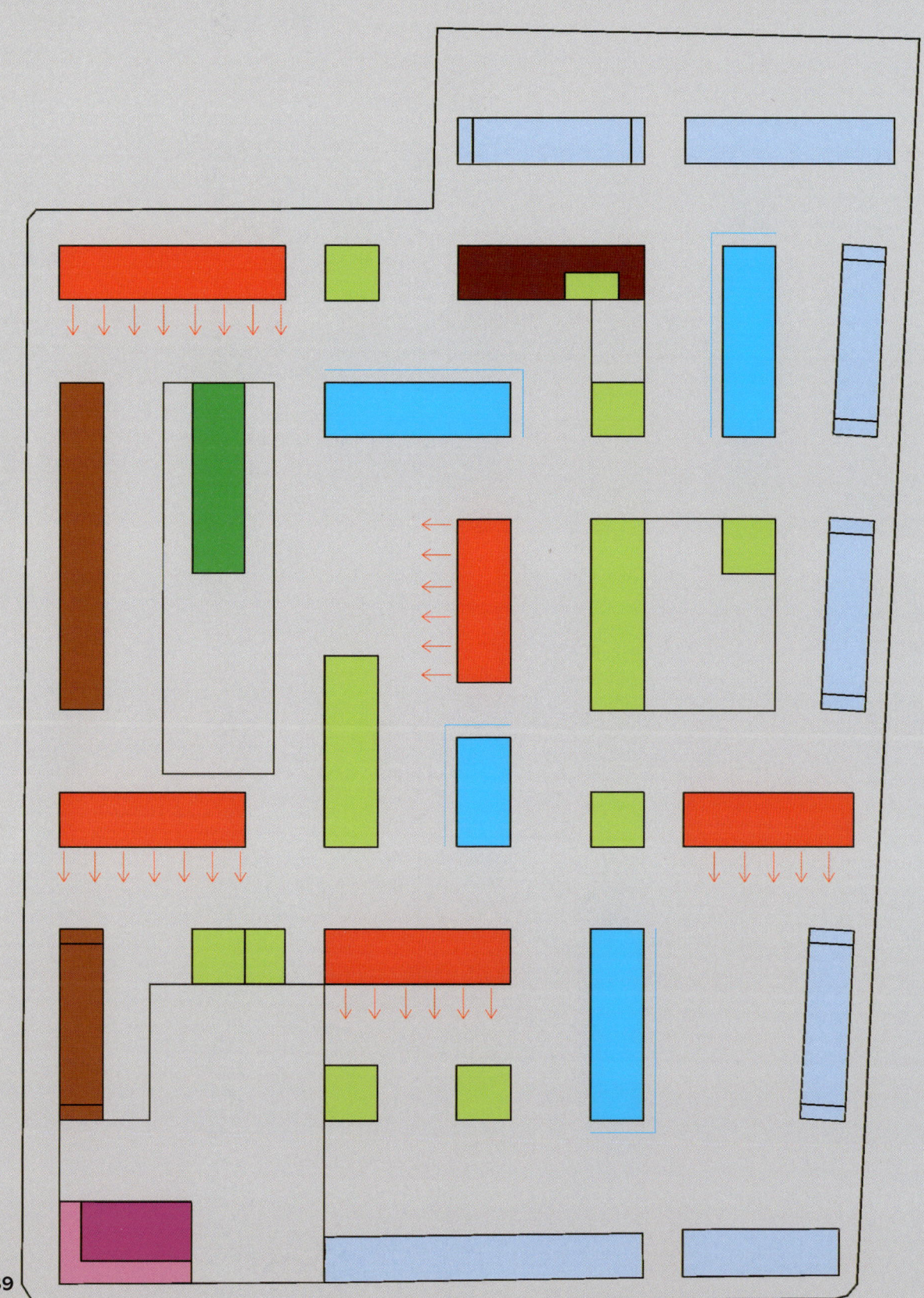

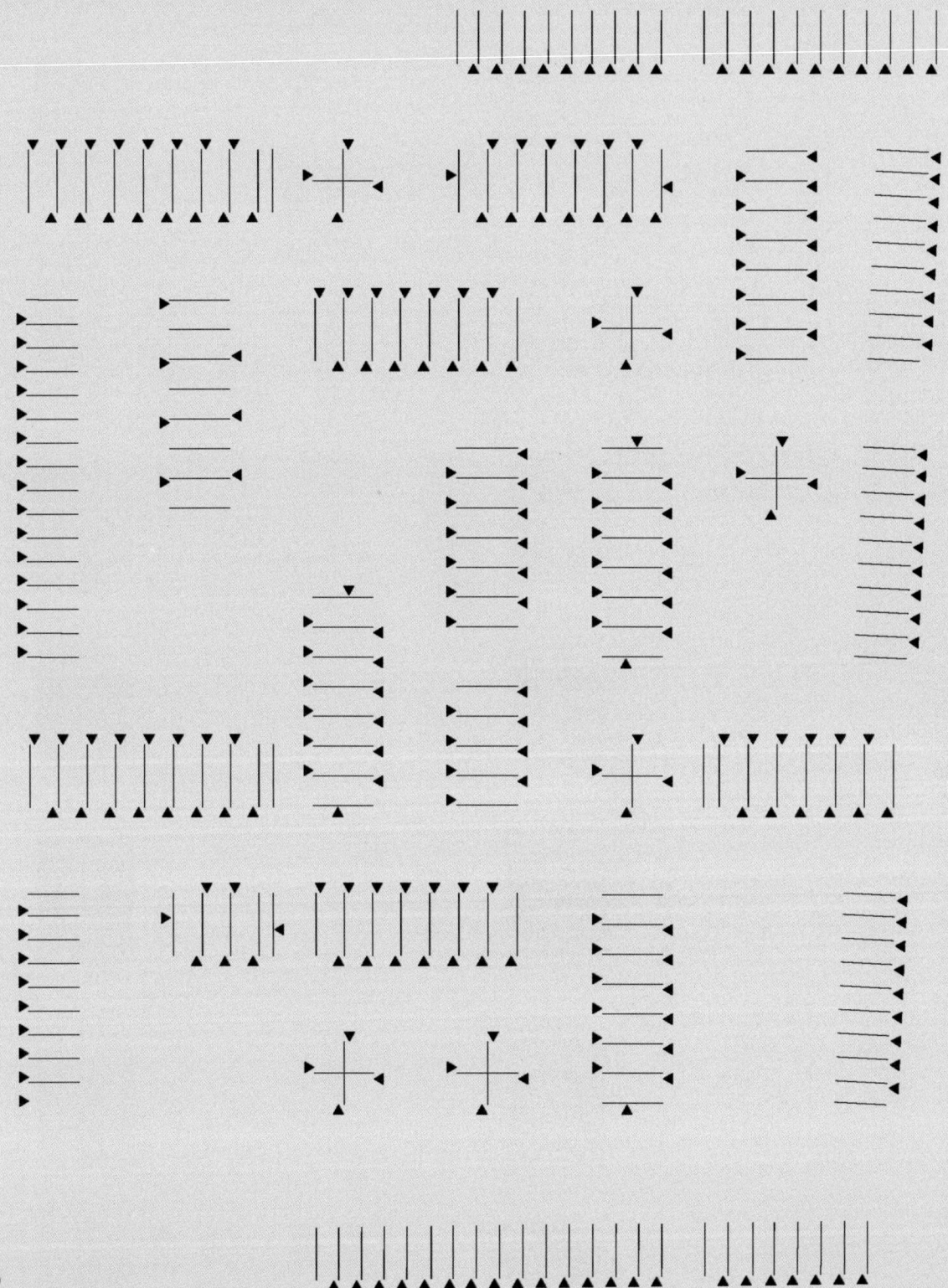

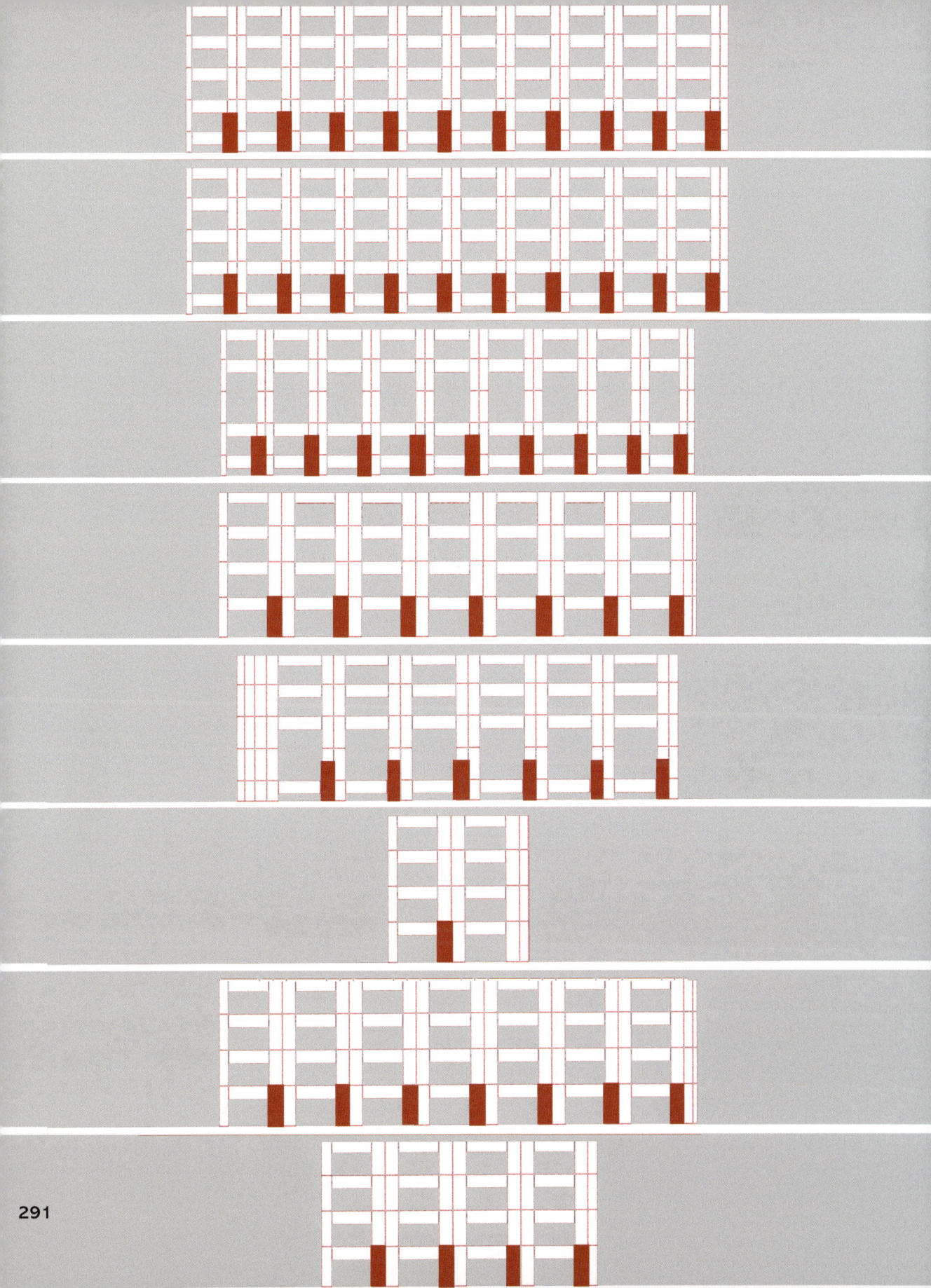

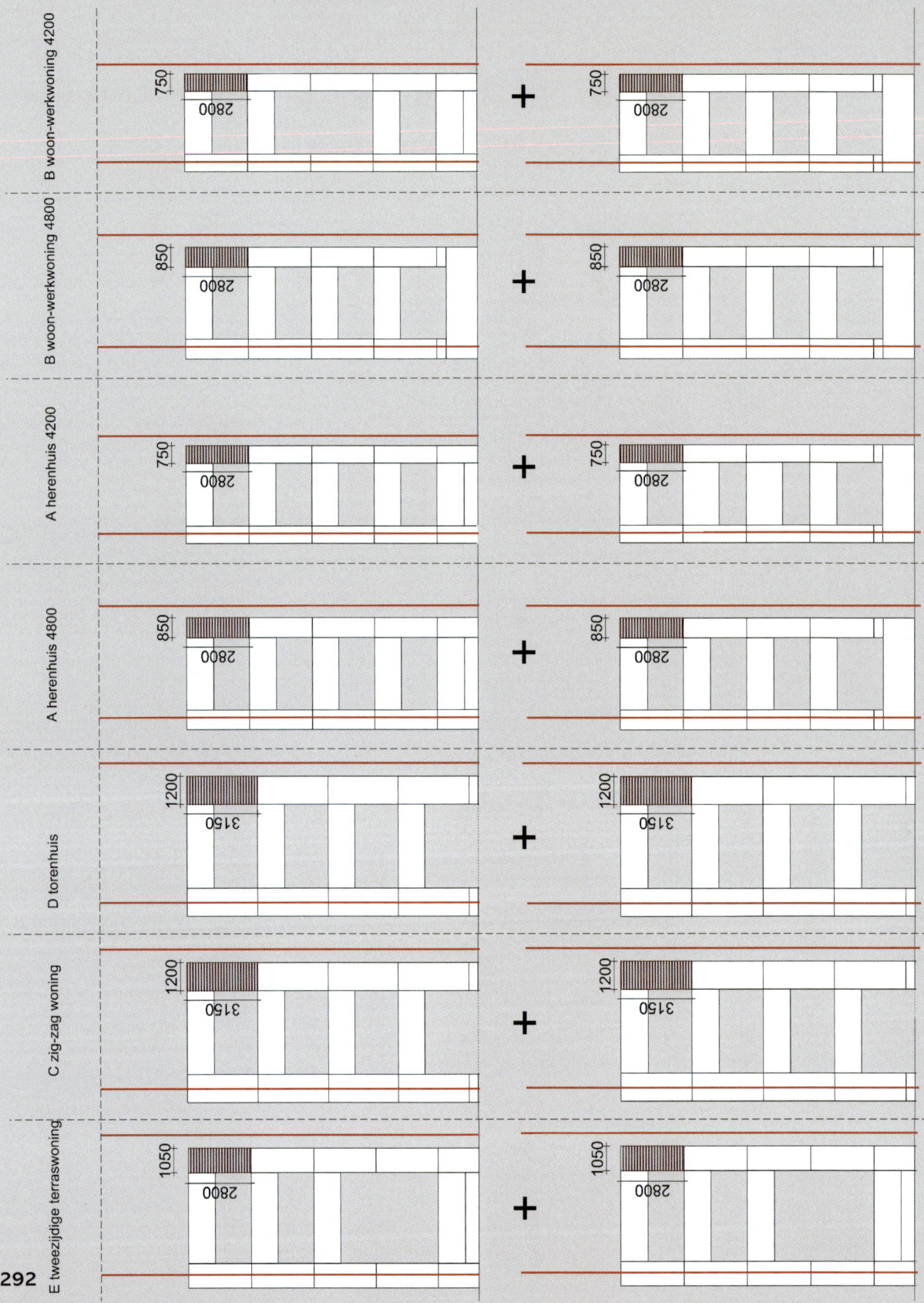

B woon-werkwoning 4200   750   2800   +

B woon-werkwoning 4800   850   2800   +

A herenhuis 4200   750   2800   +

A herenhuis 4800   850   2800   +

D torenhuis   1200   3150   +

C zig-zag woning   1200   3150   +

E tweezijdige terraswoning   1050   2800   +

292

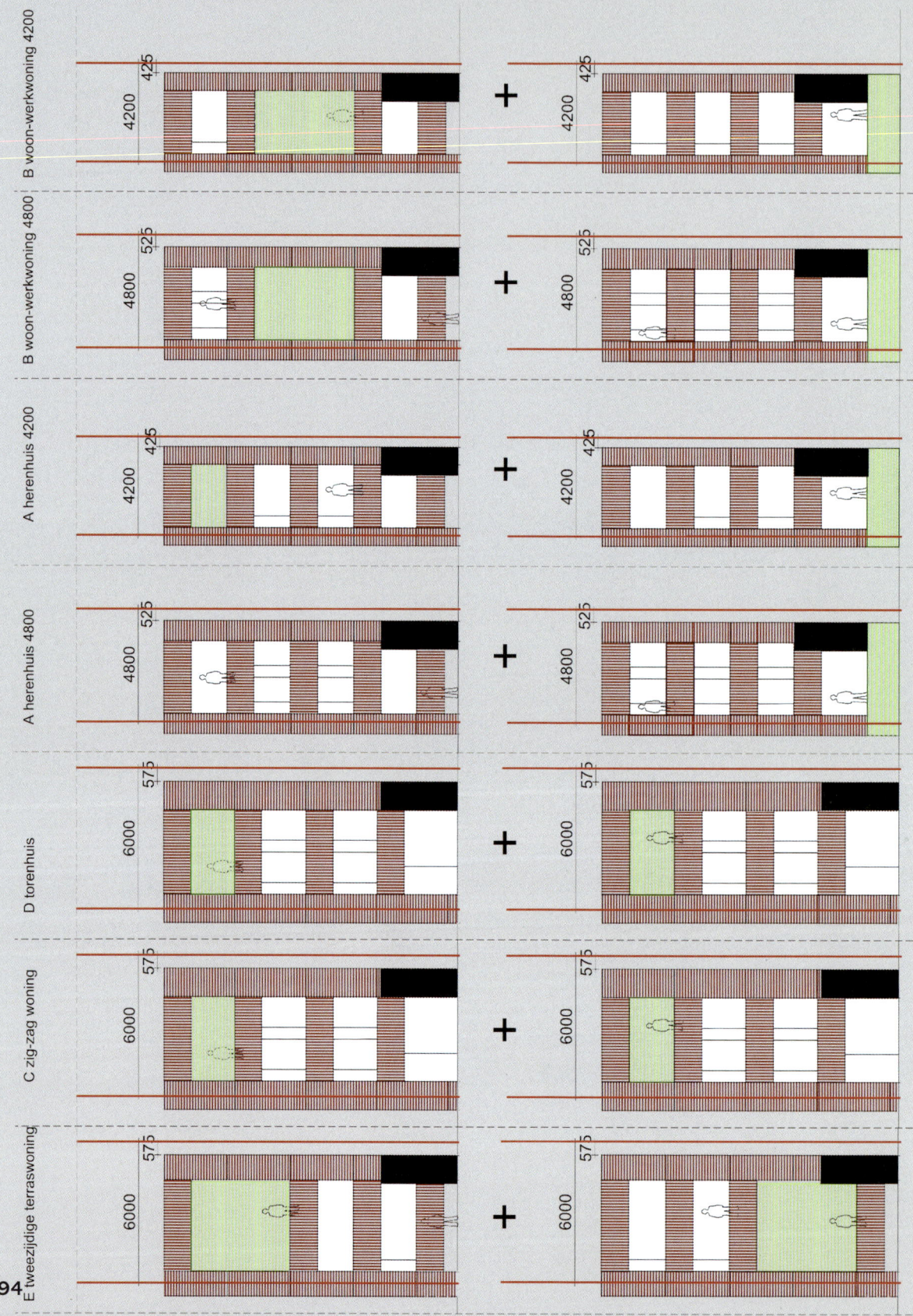

B woon-werkwoning 4200

B woon-werkwoning 4800

A herenhuis 4200

A herenhuis 4800

D torenhuis

C zig-zag woning

E tweezijdige terraswoning

294

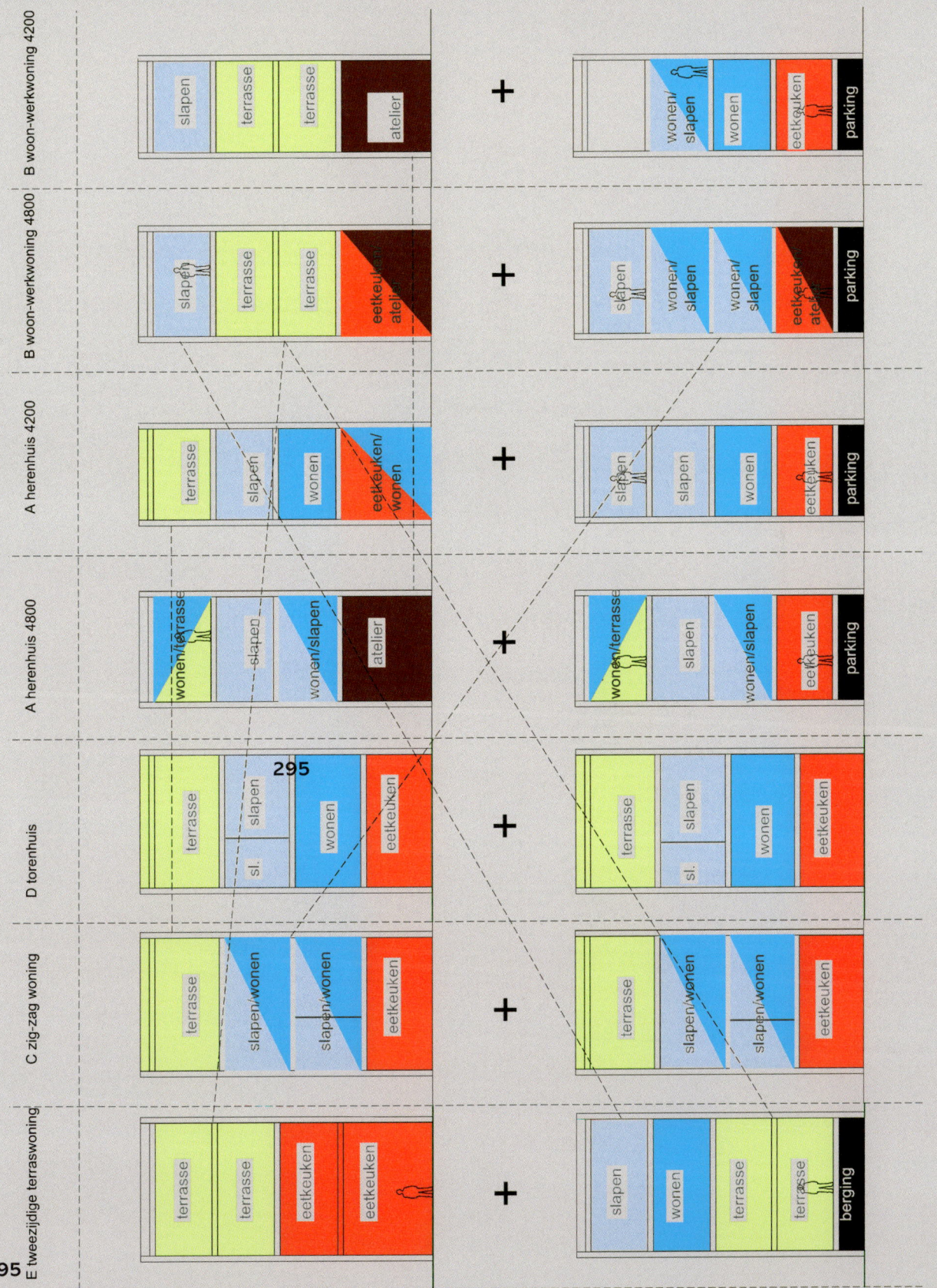

Exclusivity

Exclusivity has two different meanings: on the one hand it refers to exclusion; on the other, it denotes being unique, rarity or scarcity. Exclusive in this second sense is something that is not available in generous quantity and is therefore desirable. This latter form of exclusivity has a relevance for my architecture. No matter how absolute it might sound, exclusivity is relative. Using exclusivity precisely and purposefully does not lend more relief to just the unusual, but also to the normal.

Exclusivity

Erasmusgaarde
Exclusivity

In the neutral field of residential construction in the Erasmusgaarde in the Zuidwest district of The Hague, there are a few special spots that are marked out by the mature trees (already) standing there. At these places the ground plane is transformed into a carpet. • Within these rectangular carpets, the square and the facades are constructed from the same material. On the square the natural stone has been laid in a raw form; on the facades the same material was used in a hewn, smooth form. The large chestnut tree that stands here is floodlit in the evenings. • In the garden there are houses with glazed facades that reflect the trees and absorb the garden. • On the platform the ground level is raised. The plinth is of freestone and continues all around the blocks. • Lastly, the square is also exceptional because of the tower, which is itself exclusive, as the only vertical element in the development. The tower has two faces: an 'exterior' of glass, which bears no relation to the field, and an 'interior' that was realized in the same material as the square and has a structure that is related to that of the square.

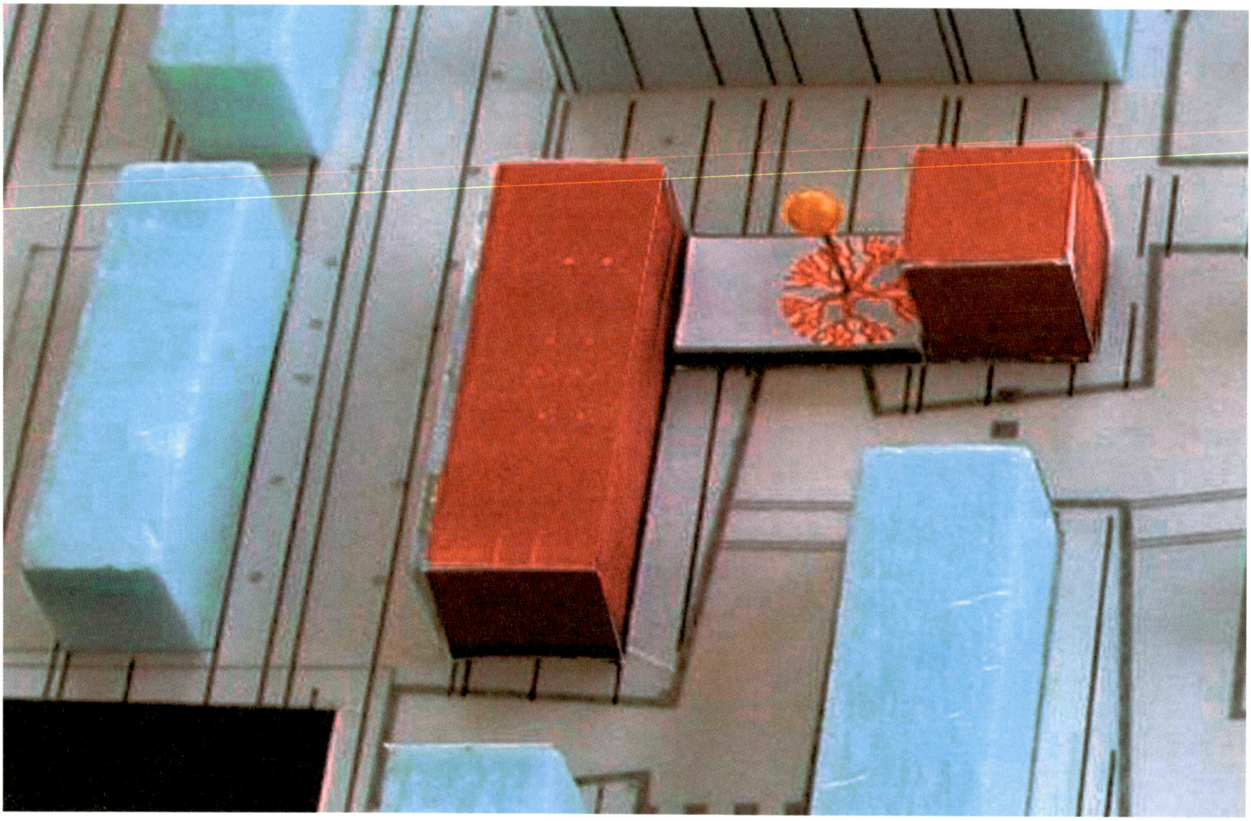

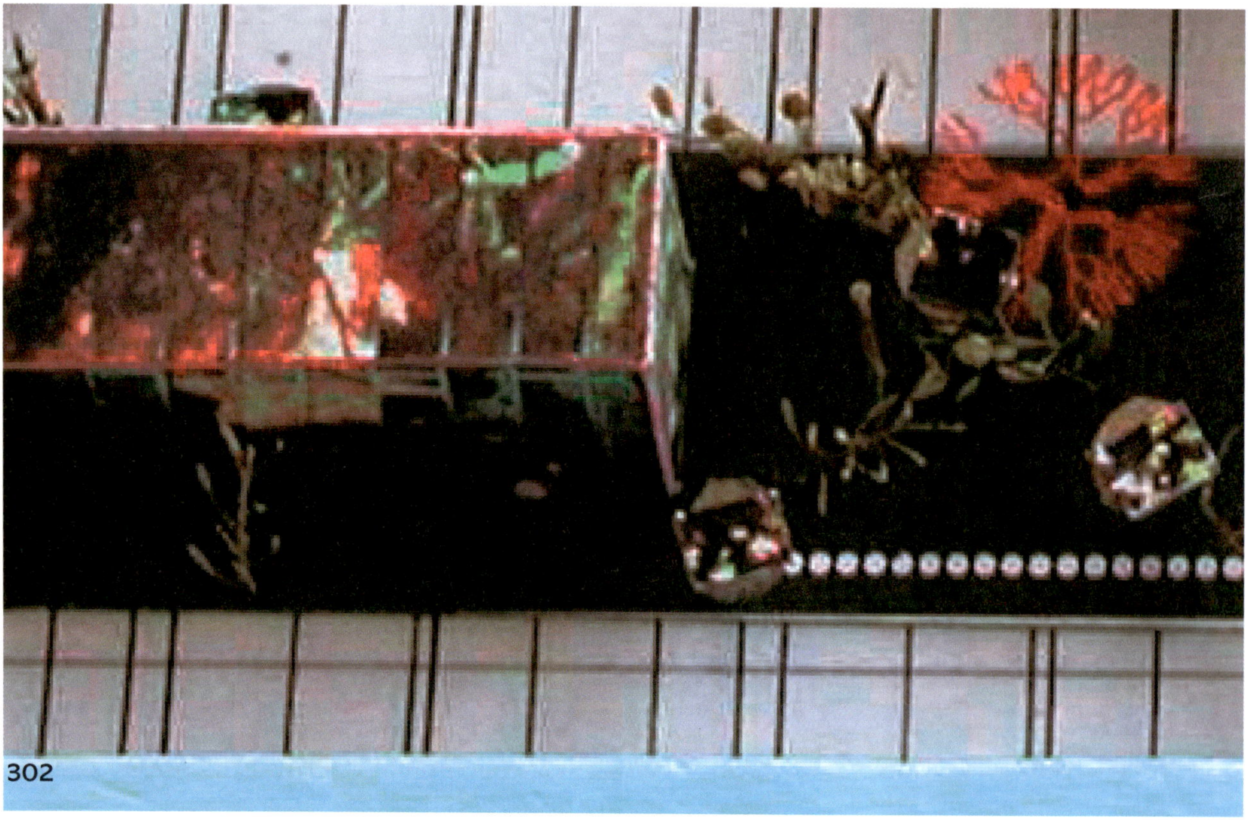

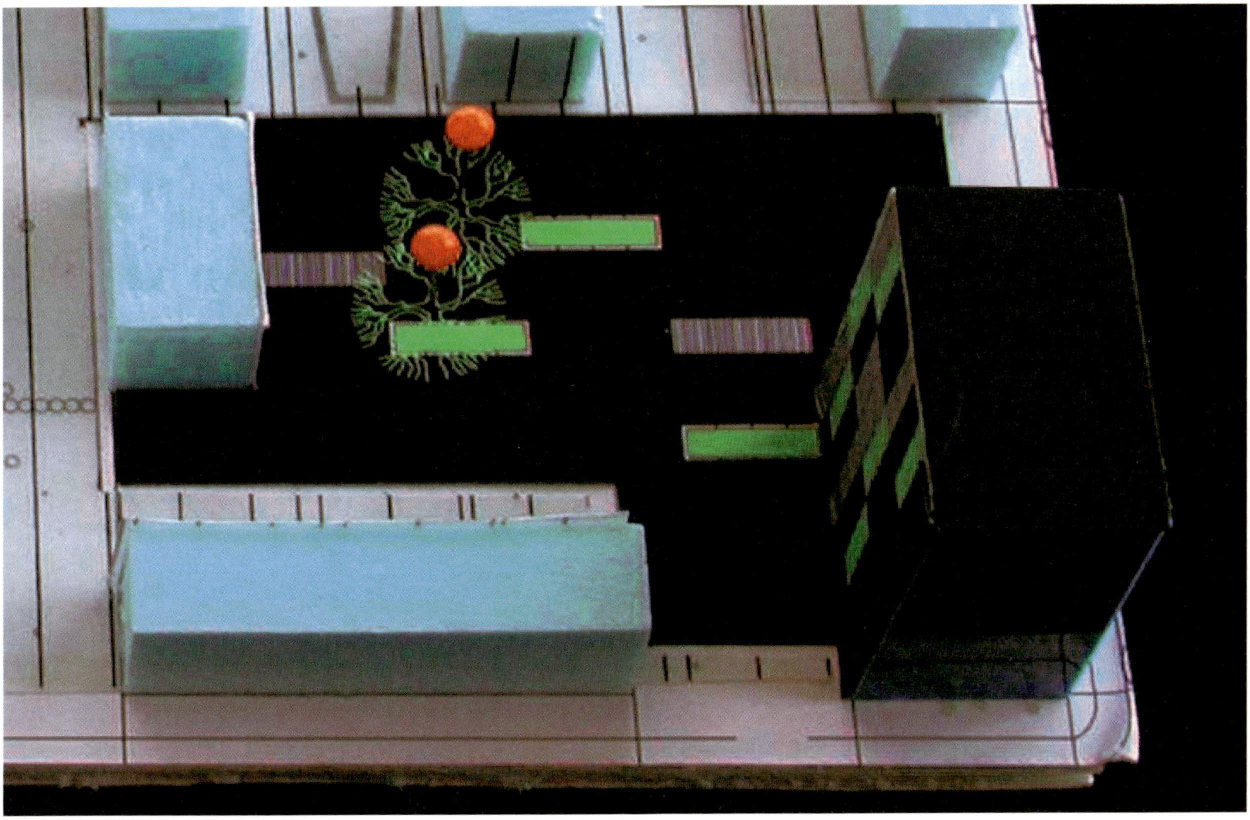

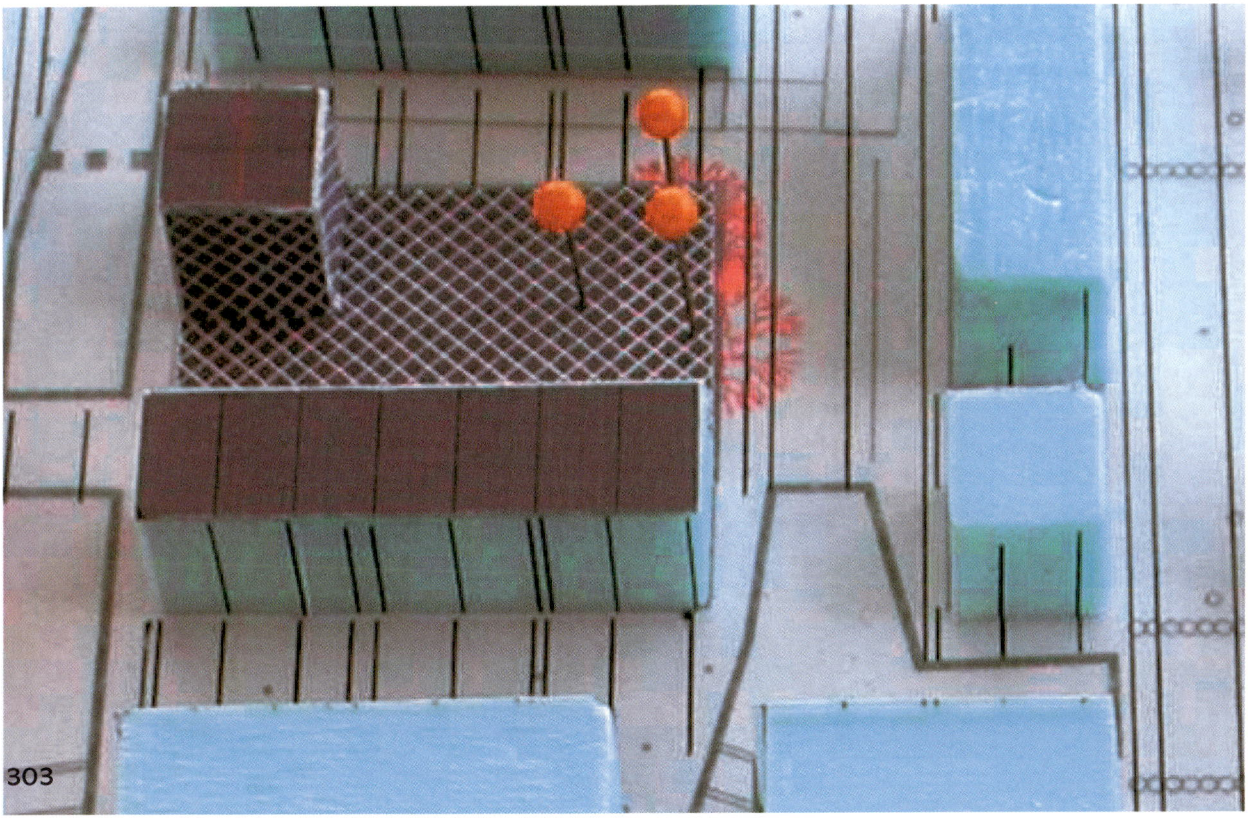

Ichthus
Exclusivity

It is primarily above the construction height of 18 metres, which was a criterion for the height of reconstruction after the Second World War, that the building next to the Laurenskerk in Rotterdam displays its exclusive character. Above this height there are apartments and penthouses that are the most spatially exclusive and, moreover, enjoy the most exclusive view – of the city but also across the 25–metre–high bamboo garden that cuts across the building.

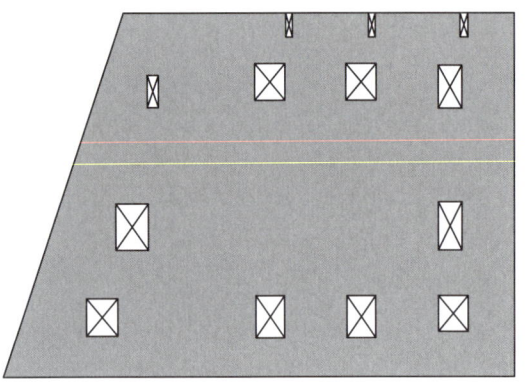

verdieping -1/-2

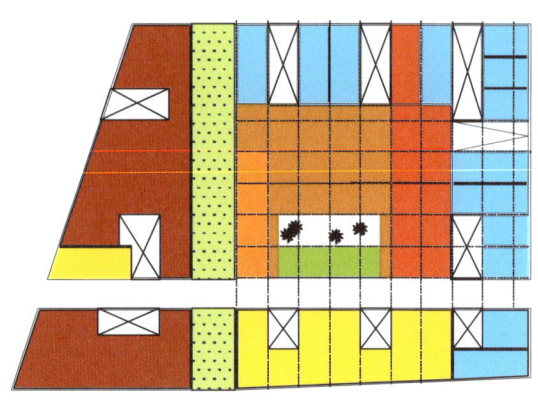

begane grond

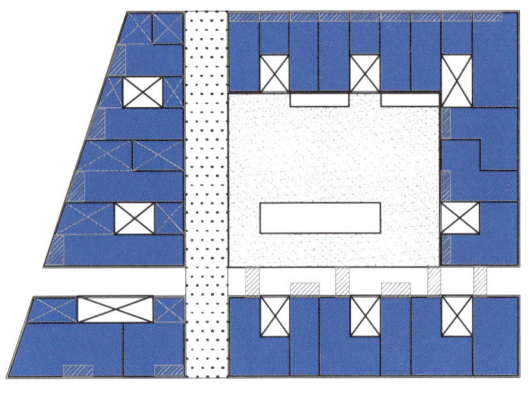

verdieping 3

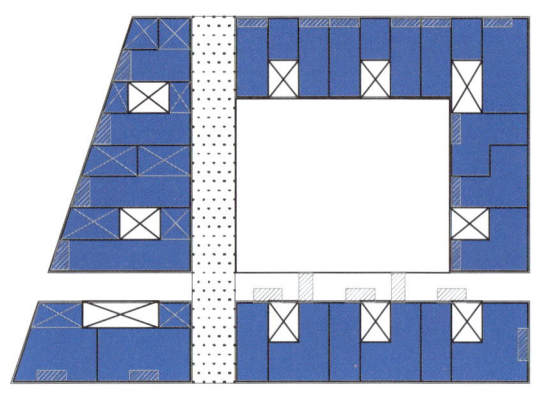

verdieping 4

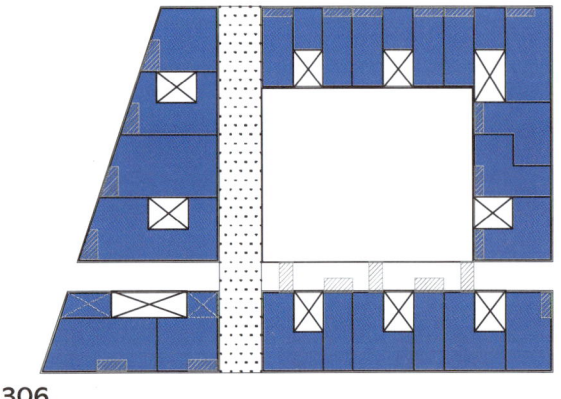

verdieping 7

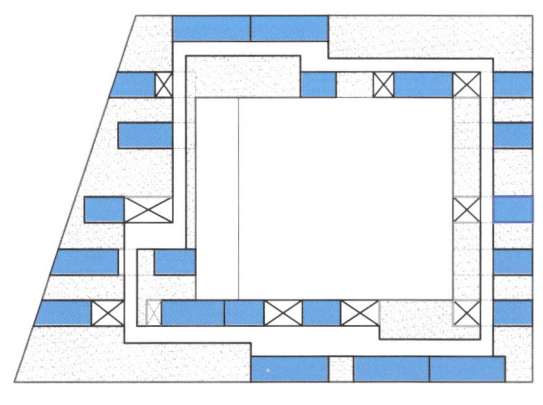

verdieping 8

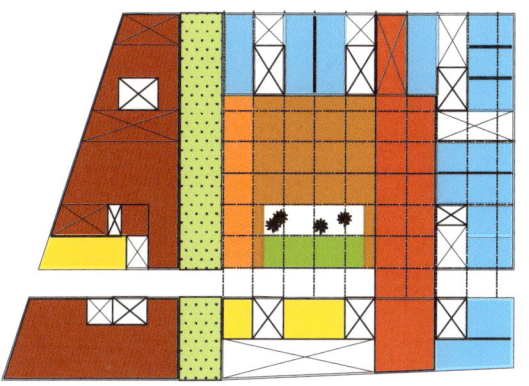

verdieping 1

verdieping 2

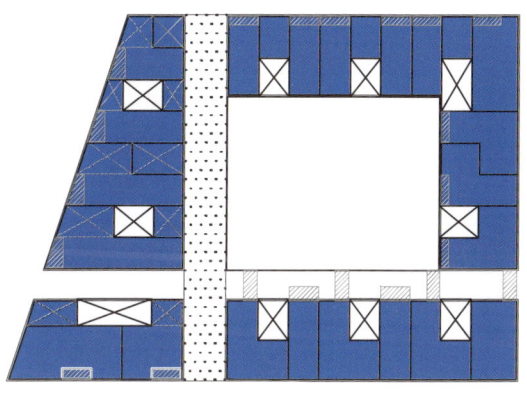

verdieping 5

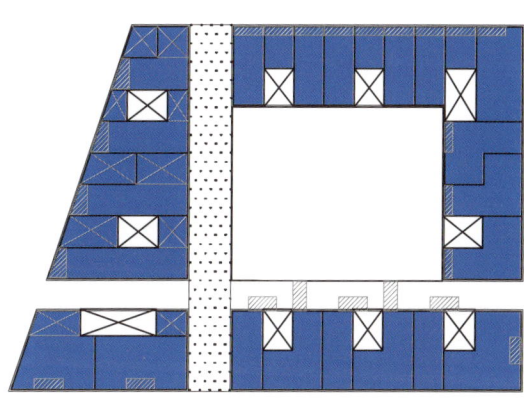

verdieping 6

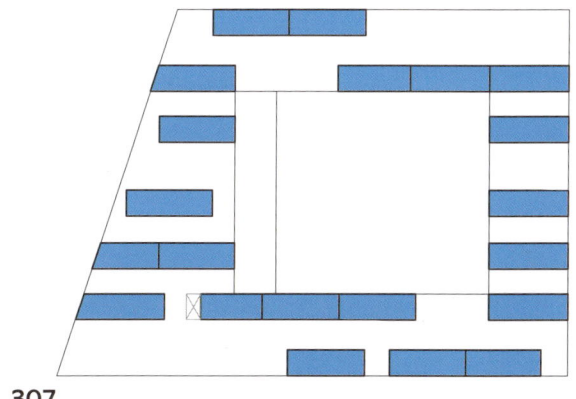

verdieping 9

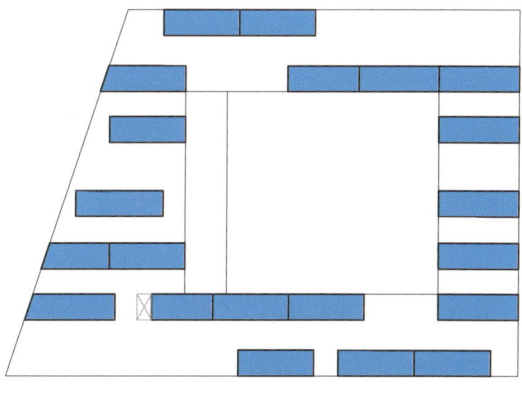

verdieping 10

307

Waterfront Exclusivity

The residential building overlooking the Westerschelde estuary, on the edge of a newly developed district of Bergen op Zoom, consists of six non–identical layers. The differences between the dwellings lend each one a certain exclusivity. The most exclusive unit on each floor is situated at the top of the building: a transparent penthouse that enjoys the water and vista to the full.

0

10m

Block 30
Exclusivity

Eight compact buildings, each containing five housing units, were designed for Block 30 in Amsterdam's IJburg expansion project. In Block 30, for which Liesbeth van der Pol was the coordinating architect, they are eight exceptions. The sharply angled corners emphasize the autonomy of these buildings. The entrance to the units is also distinctive. The staircase is double height and natural light enters from behind. The character of the dwellings is determined by all the amenities being accommodated in a compact zone, making it possible to create flexible, loft–like spaces.

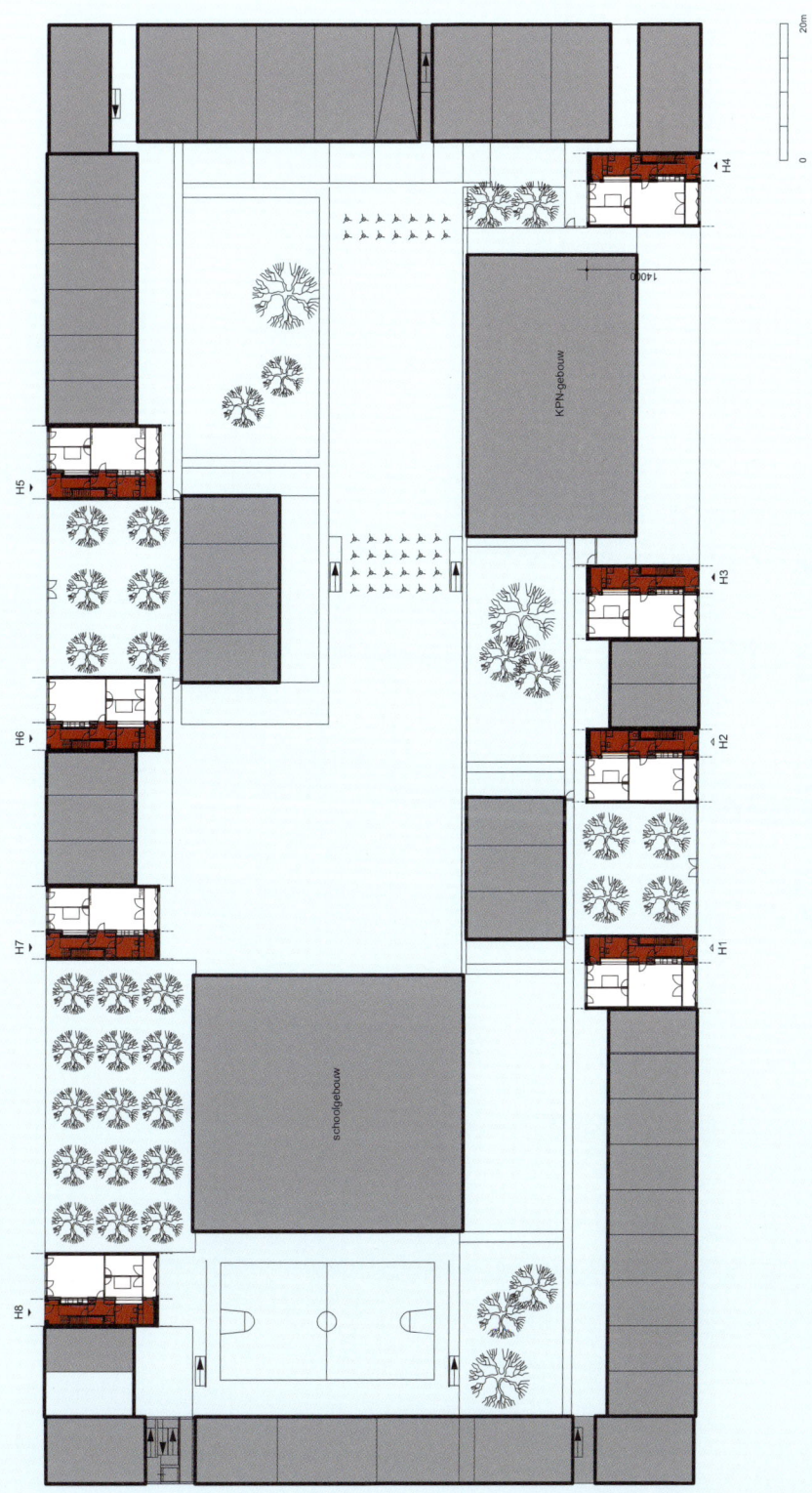

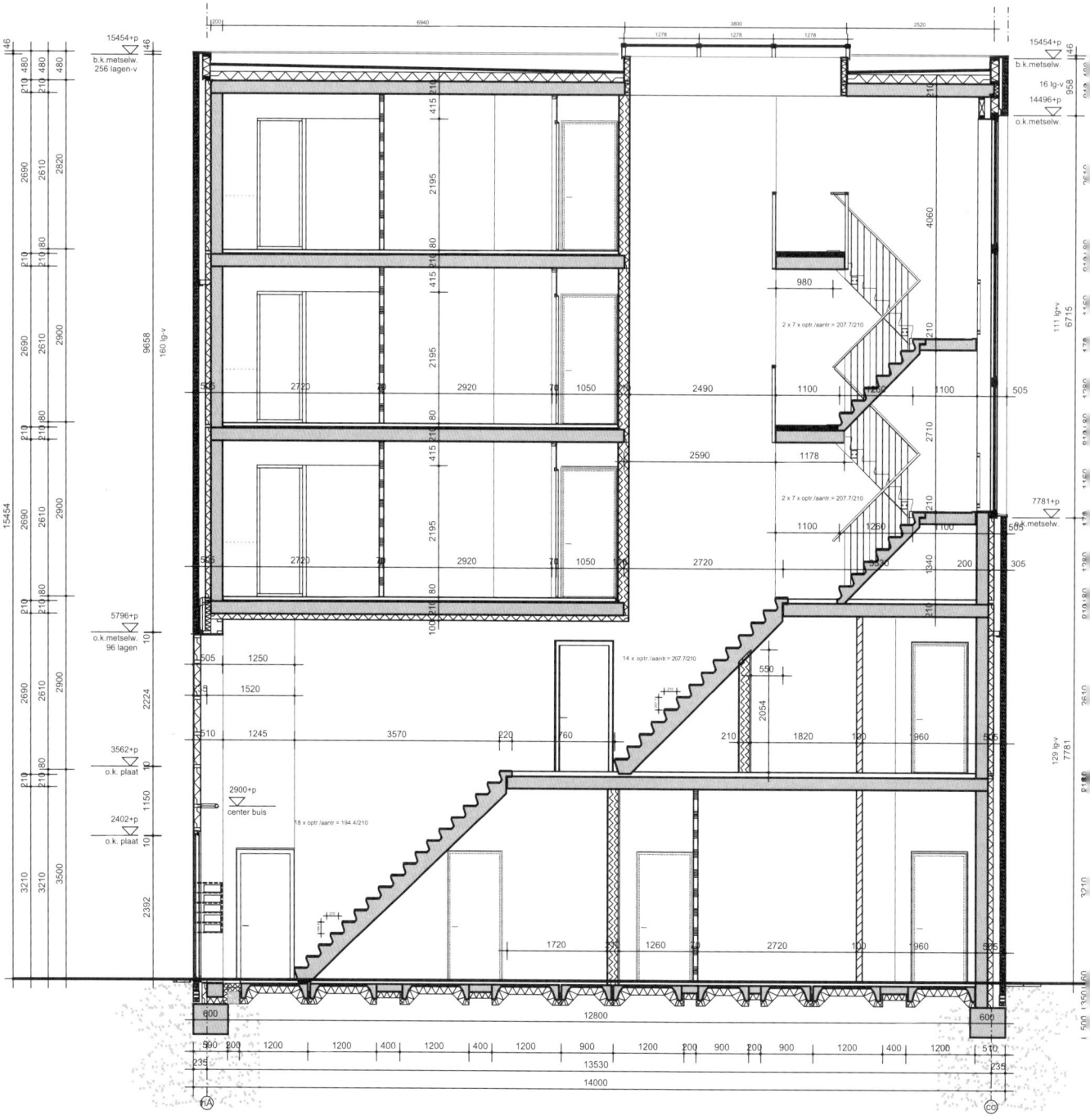

314

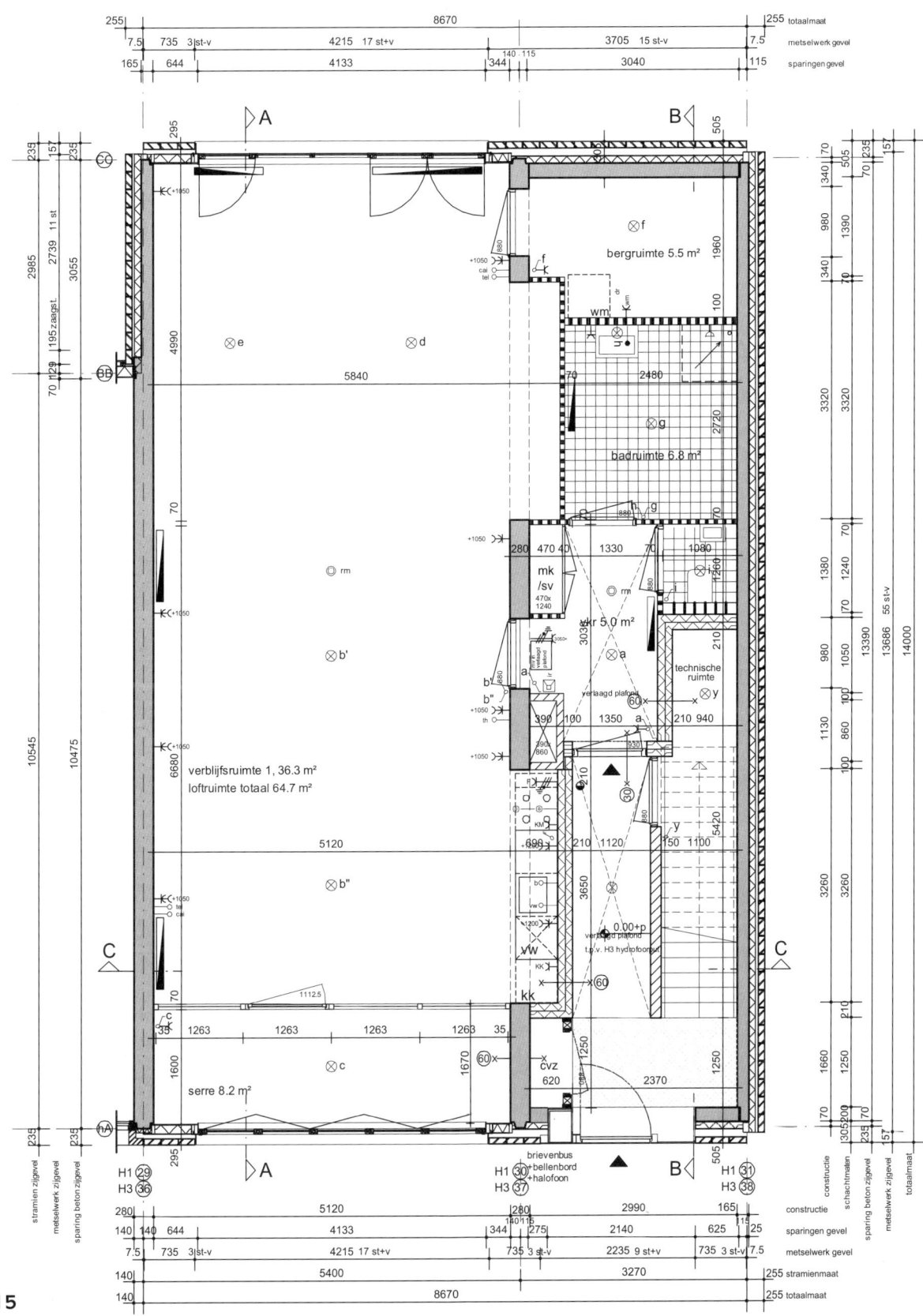

315

Suytkade
Exclusivity

The gold–coloured aluminium in the facades of this detached, omni–sided residential block in a neighbourhood in Helmond, to an urban design by Soeters Van Eldonk Ponec, underscores the prima donna character of this building and the exclusive character of the dwellings. The dwellings have open loggias and balconies that are sometimes double–height or double–width and then extend across the frontage of another unit. These unorthodox open–air spaces lend the dwellings an exclusive character.

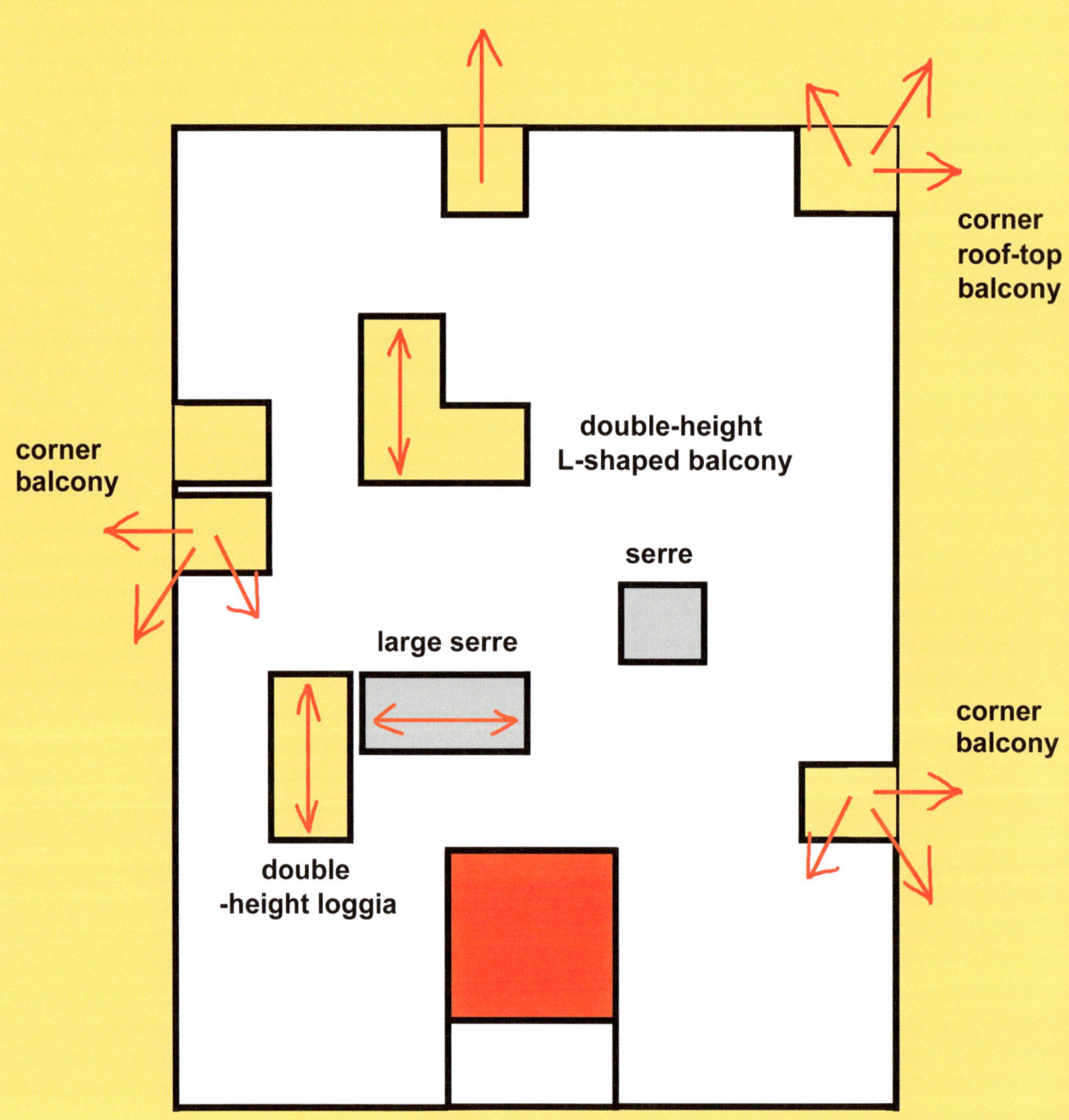

roof-top balcony

corner
roof-top
balcony

double-height
L-shaped balcony

corner
balcony

serre

large serre

double
-height loggia

corner
balcony

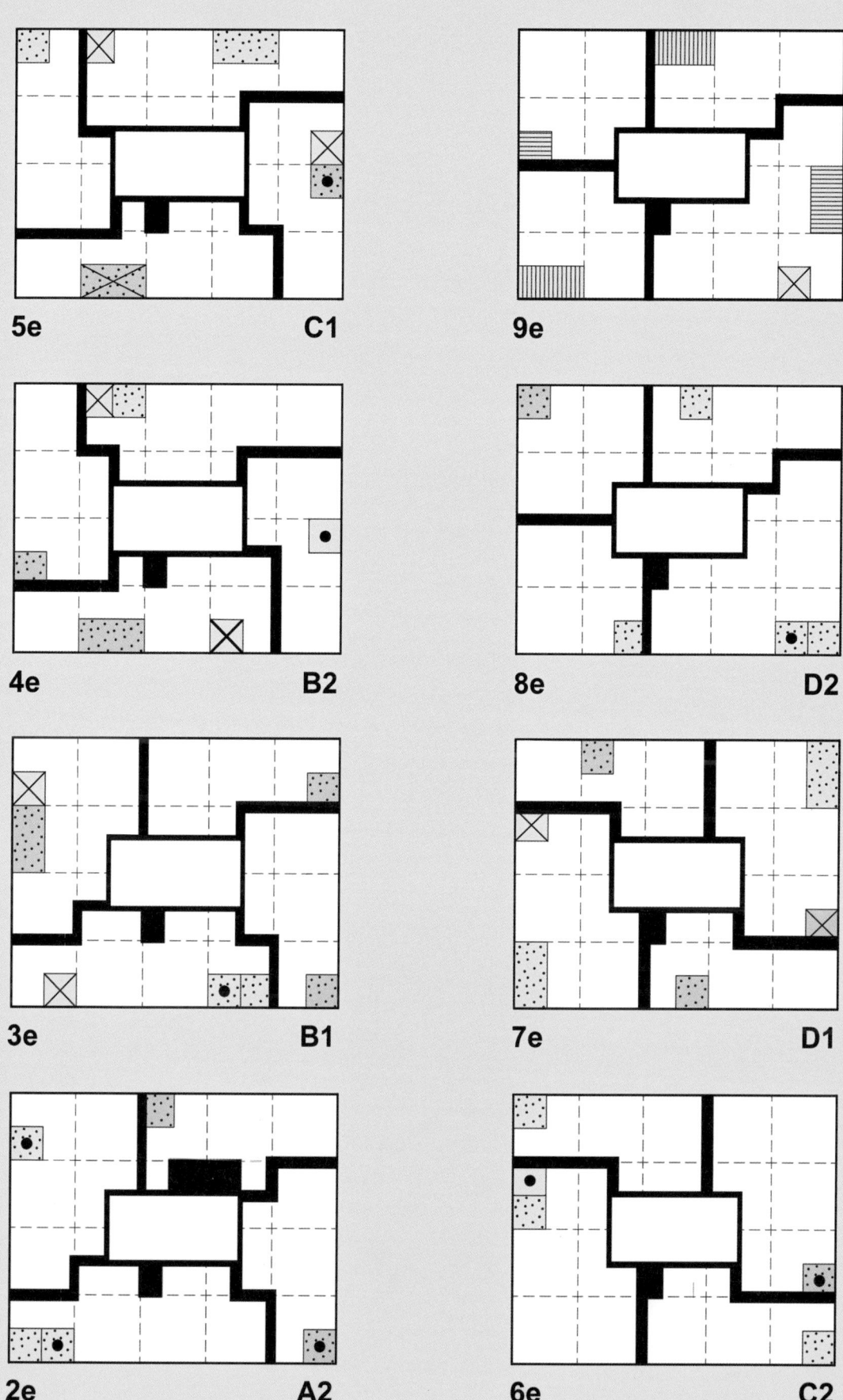

5e   **C1**

9e

4e   **B2**

8e   **D2**

3e   **B1**

7e   **D1**

319

2e   **A2**

6e   **C2**

Pendrecht
Exclusivity

A tower was planned within the design for the transformation of an area on the south side of the Rotterdam district of Pendrecht. The original plan by Lotte Stam–Beese from the 1950s provided a logical anchor for the placement of this building. The tower stands at the crossing–point of the axes in the centre of the district, the point where all the special elements converge: the Plein 1953 square, water and greenery.

Het Funen
Exclusivity

The commission for this building with ten dwellings at Het Funen in Amsterdam was to make a 'hidden delight', a block that occupies a unique position with regard to the other blocks and is exclusive in the spatial quality and orientation of the dwellings. The dwellings are 27 metres deep and have a width that is staggered, from three to two metres, to four to three metres. This long space is pierced at various points so that light penetrates deep into the dwelling. There is a maximum contrast between this space and the openness of the facades, which wholly consist of pivoted glass doors that provide a view of the park between the blocks. The dwellings are open on two sides, so there is no front or rear.

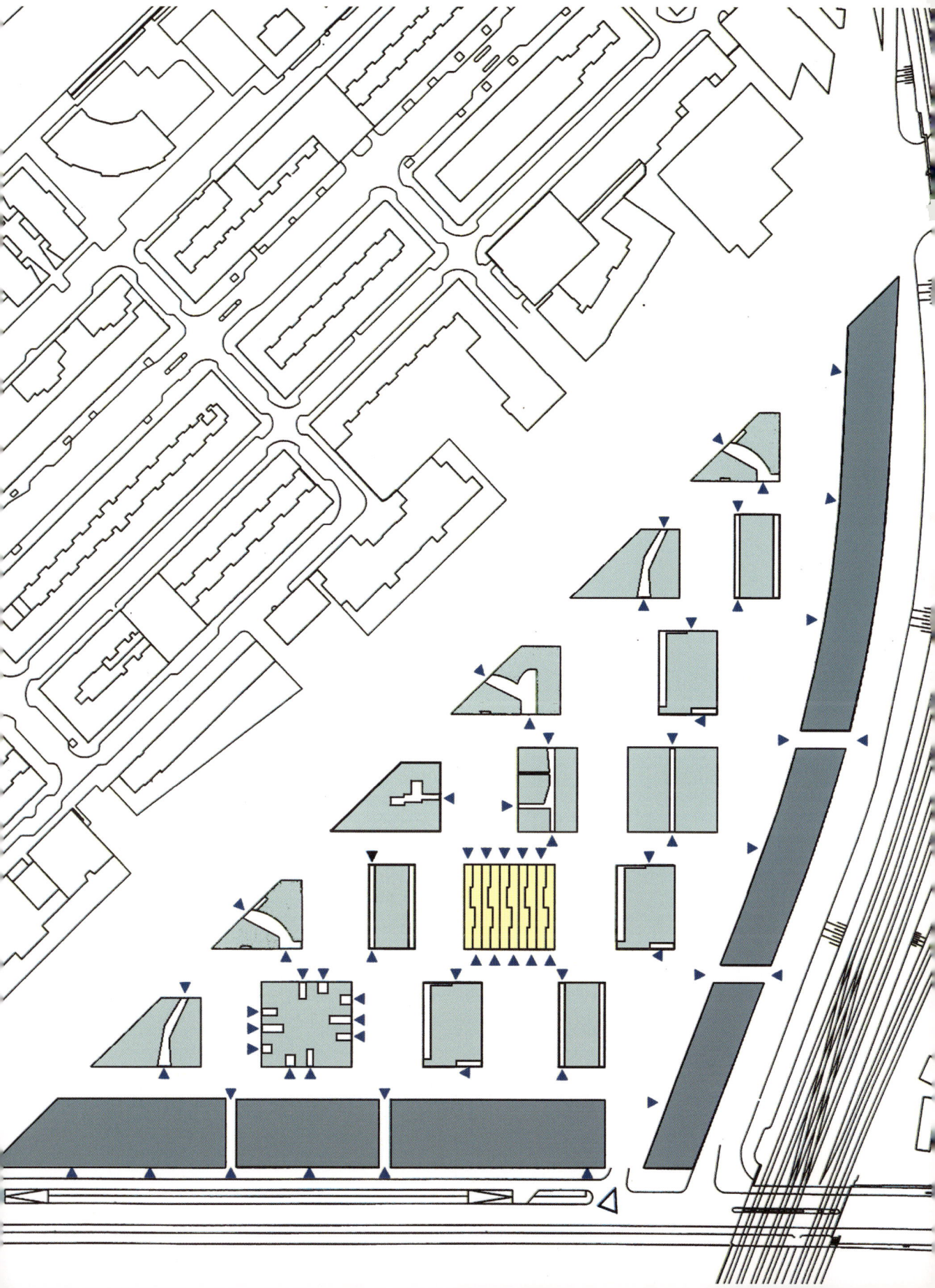

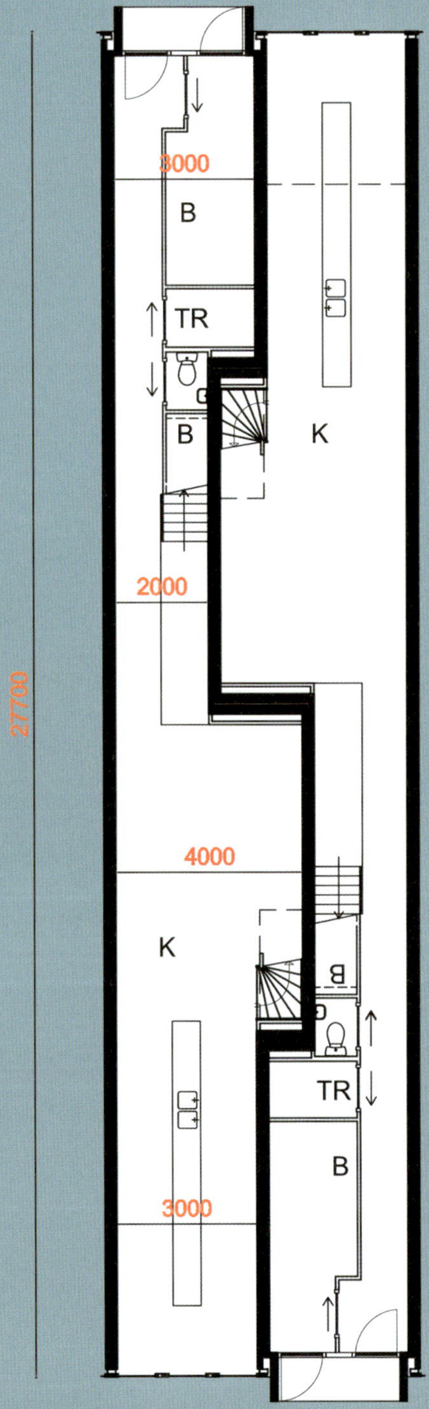
3000
B
TR
B
K
2000
27700
4000
K
B
TR
B
3000

begane grond

WK /
WERK
WK /
WERK

eerste verdieping

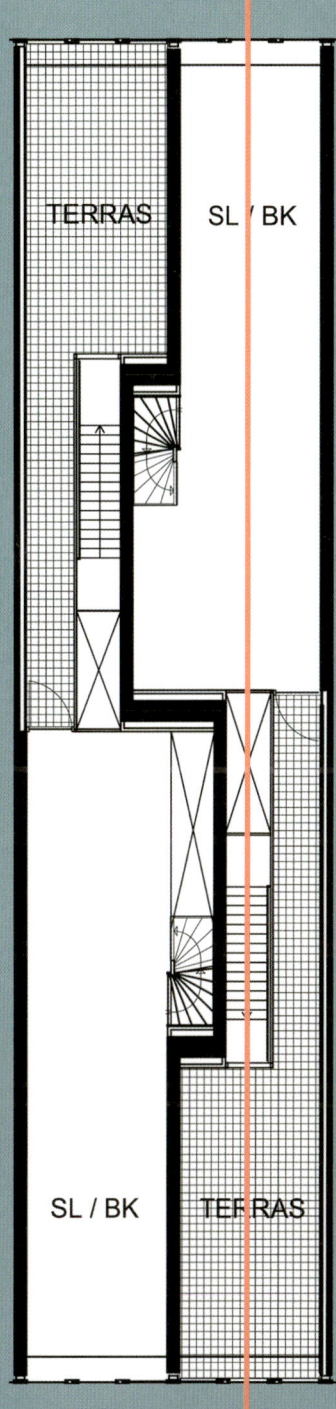

TERRAS   SL / BK

SL / BK   TERRAS

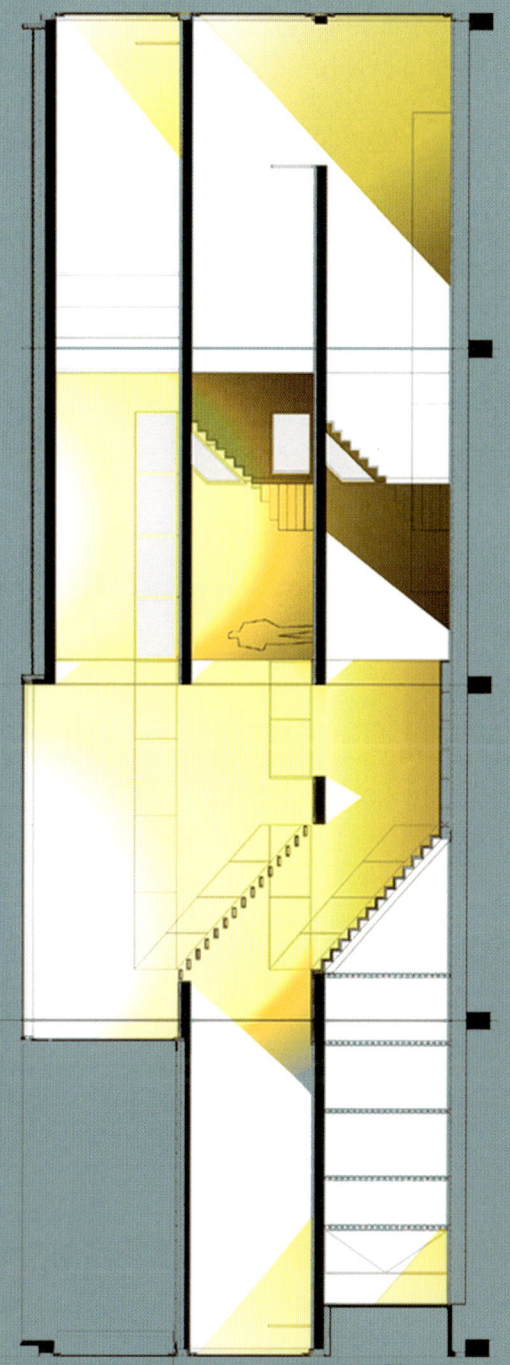

327   tweede verdieping

Neutrality

Under certain circumstances it is appropriate for architecture to be neutral: leaving it devoid of striking features, giving it no dominant orientation, no midpoint. The objective here is to leave space. • This impartial neutrality, like that of a series of parallel lines or a grid, is not the same as restraint in architectural expression, which is an aspect of my architectural signature. While all my architecture is characterized by a certain restraint, it is not all neutral.

Neutrality

Block 34
Neutrality

The essence of grids, including that of Amsterdam's IJburg expansion project, is the lack of a dominant direction. The logic of IJburg's urban plan is continued in the architecture of Block 34, where there is an equilibrium between open and closed and between flatness and relief in the facades. The neutral grid of square windows continues all round the block and adapts to the specific conditions. The neutrality is disturbed by windows that open outwards, contributing to a wealth of contrasts in which the order of the grid is always clearly present.

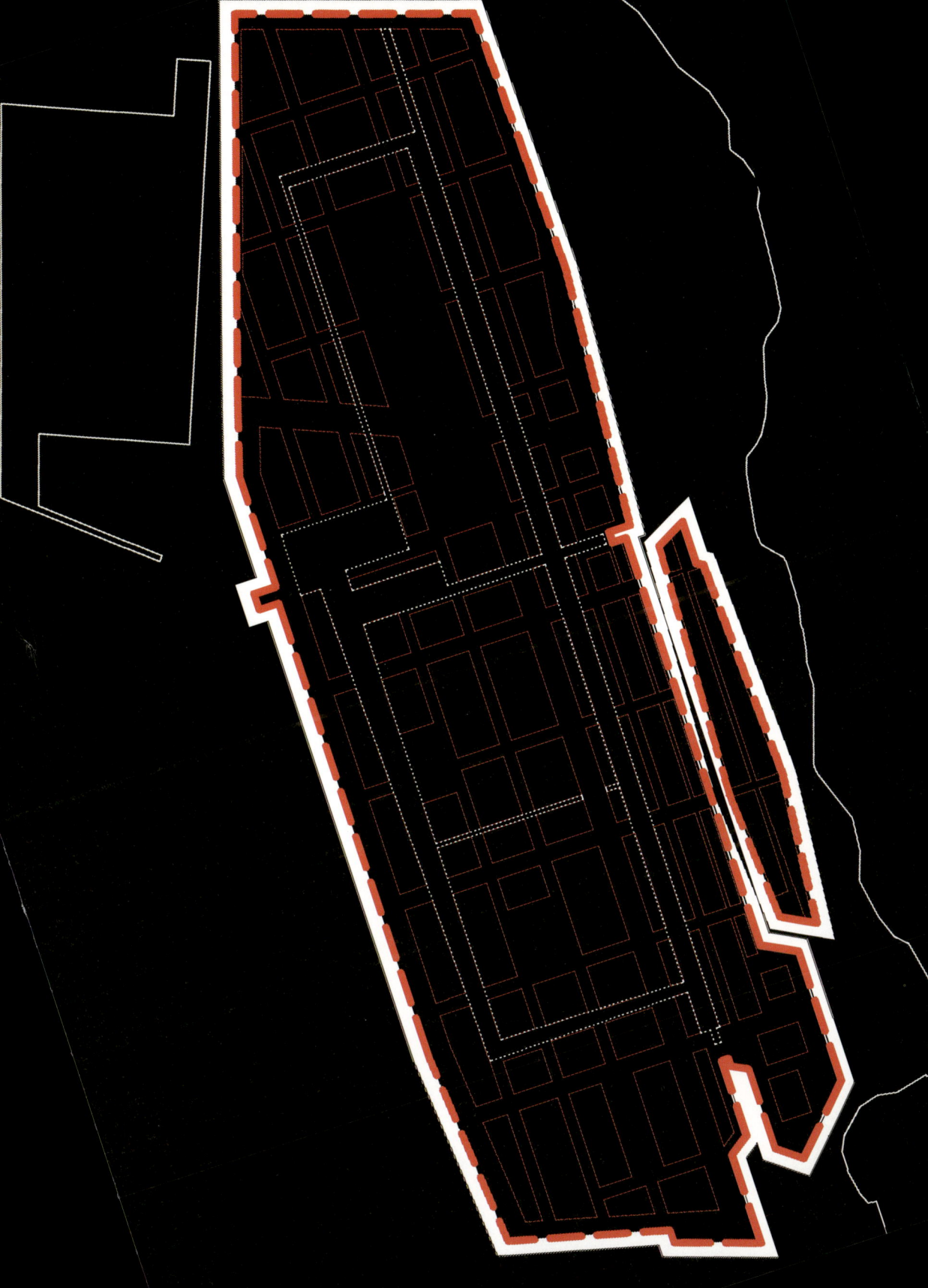

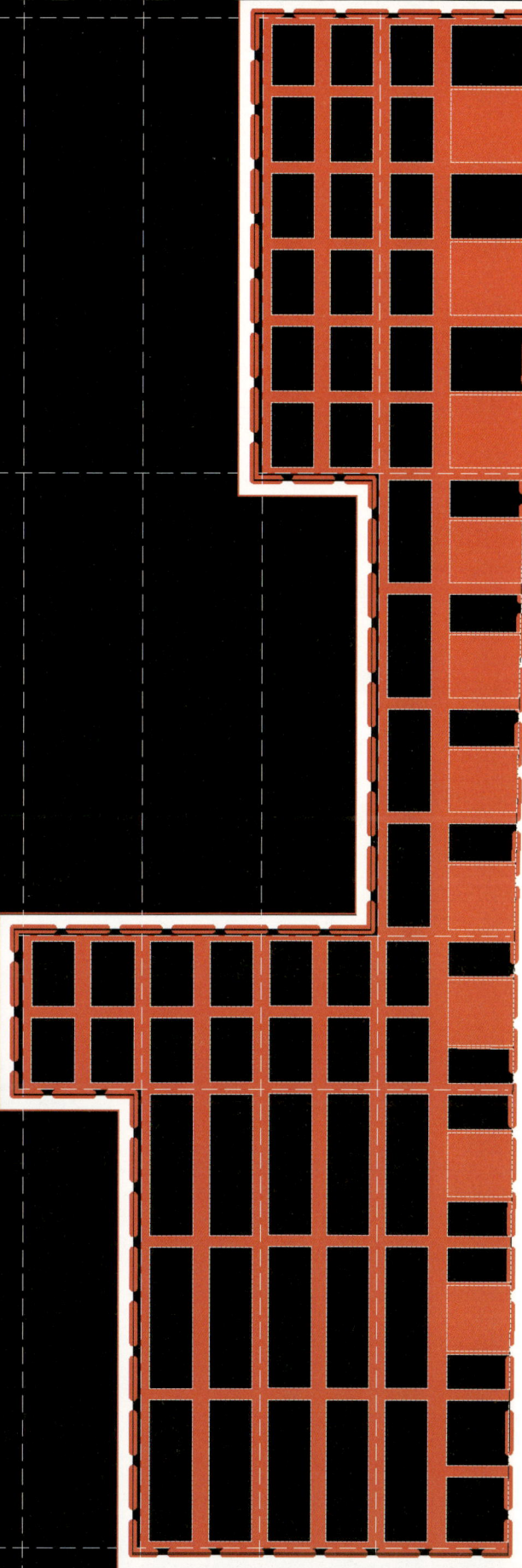

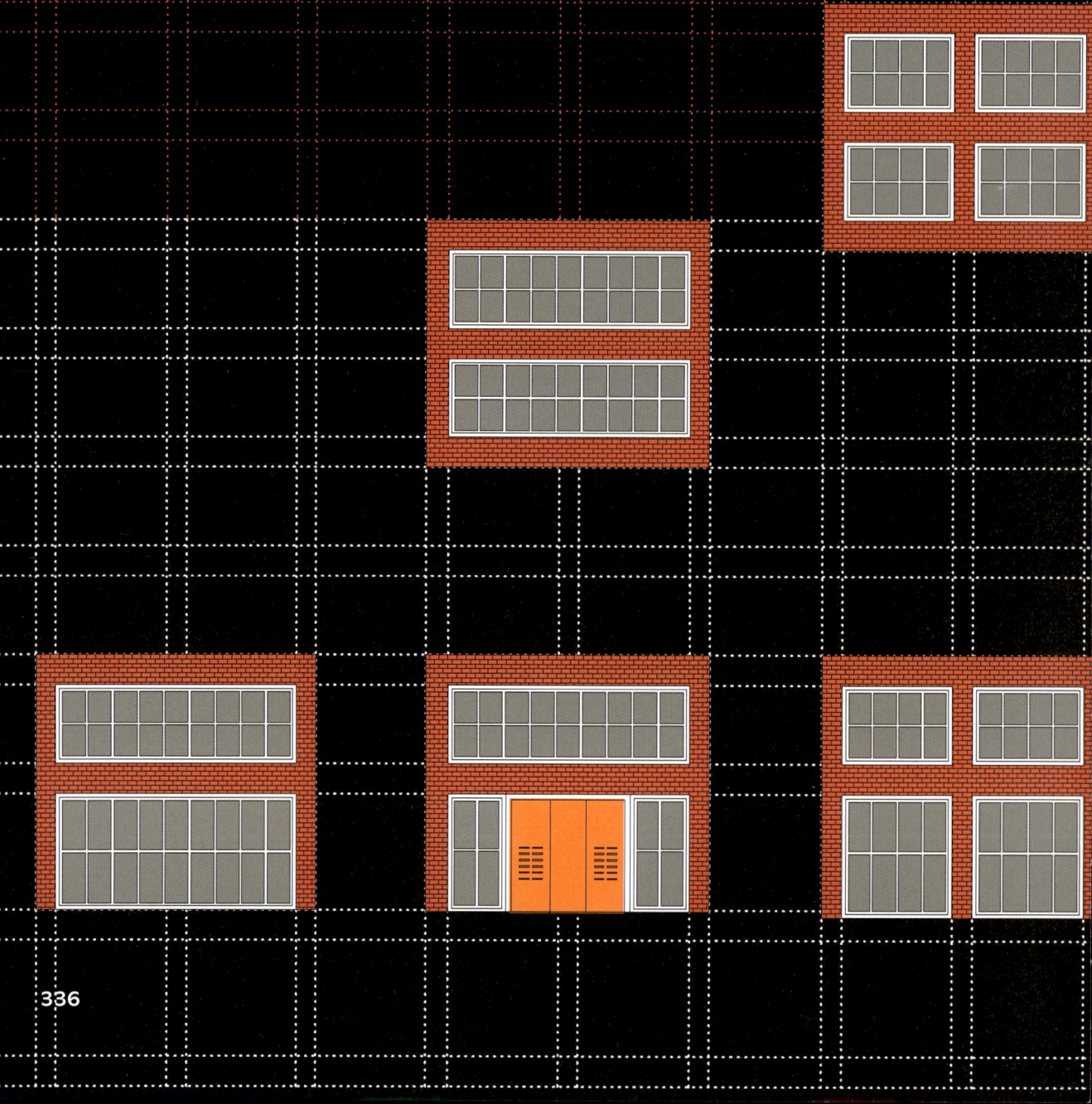

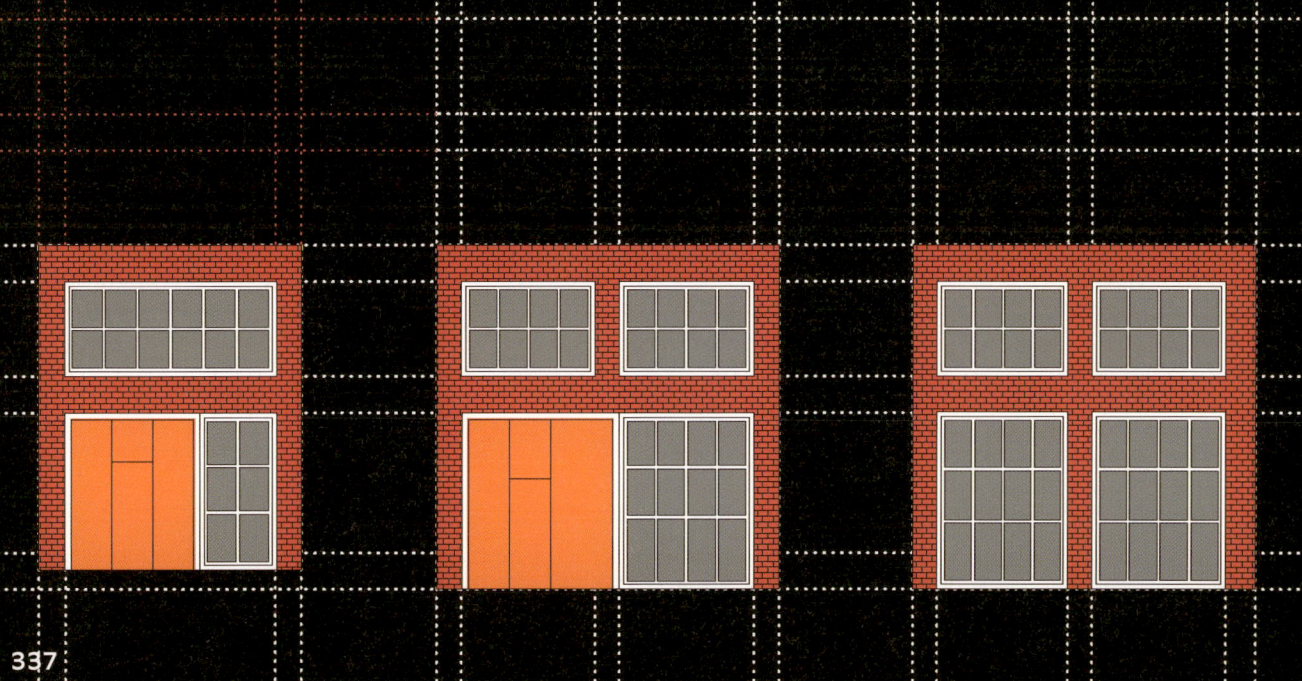

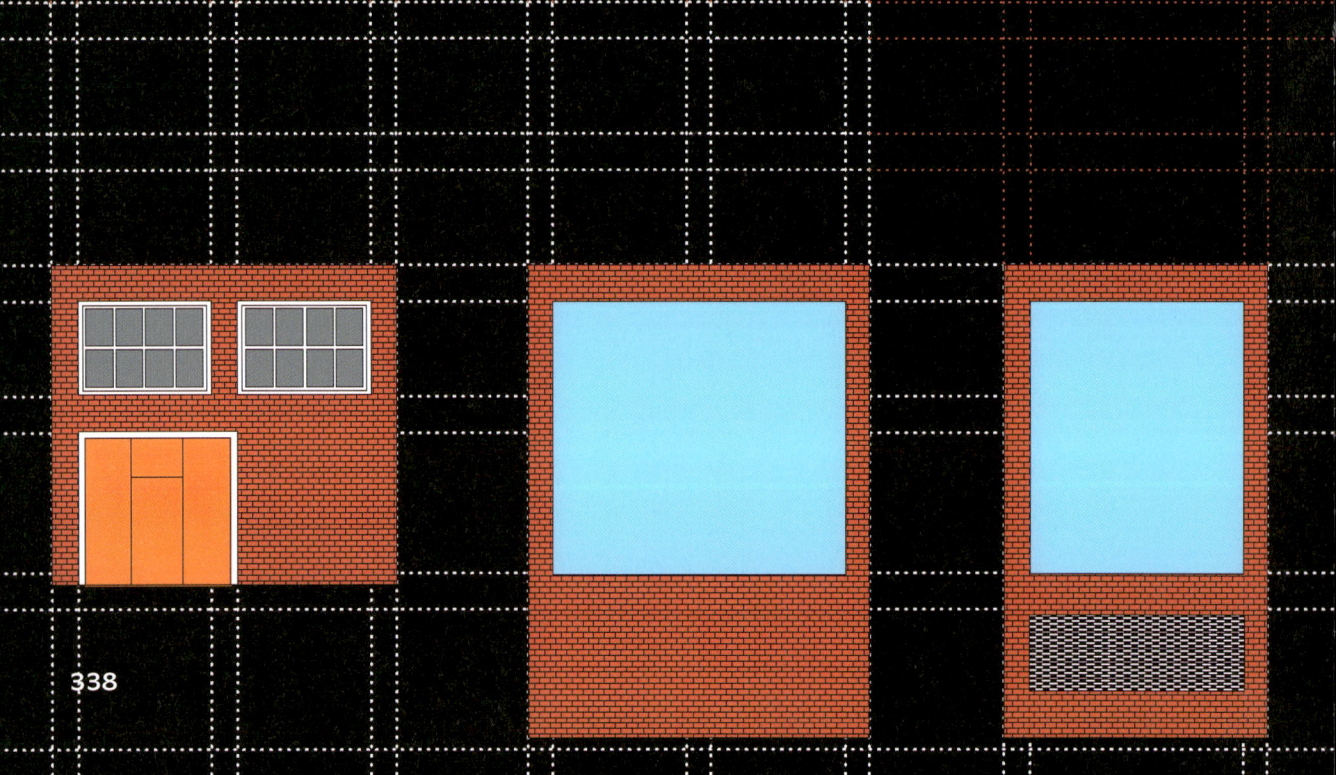

338

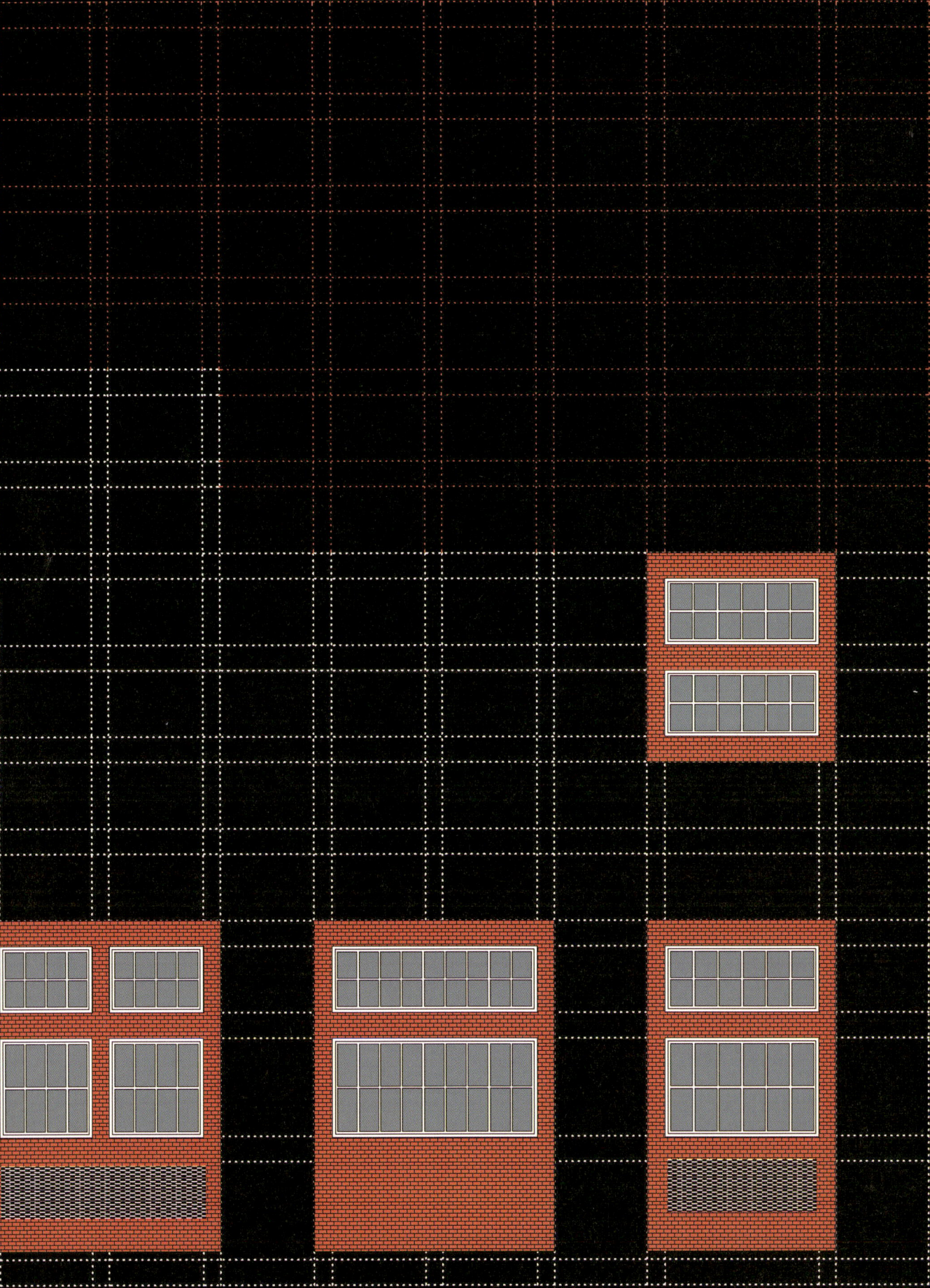

Haarlem
Neutrality

In the floor plans for the housing project in the Zuiderpolder in Haarlem, the corridor across the width of the unit and the visual corridors from front to back are of equal organizational importance. These orientations keep each other in equilibrium and there is no strict spatial hierarchy within the dwellings.

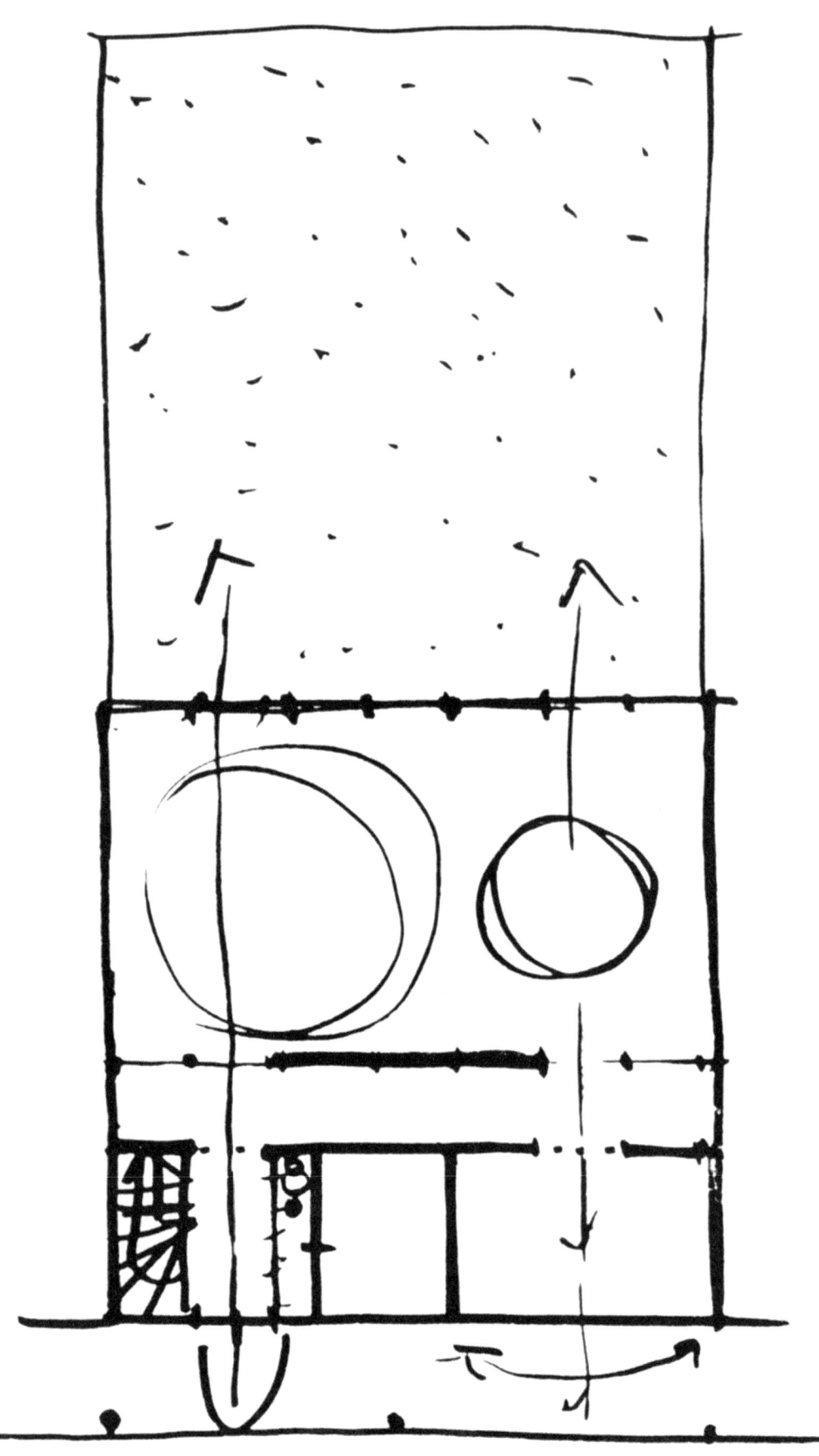

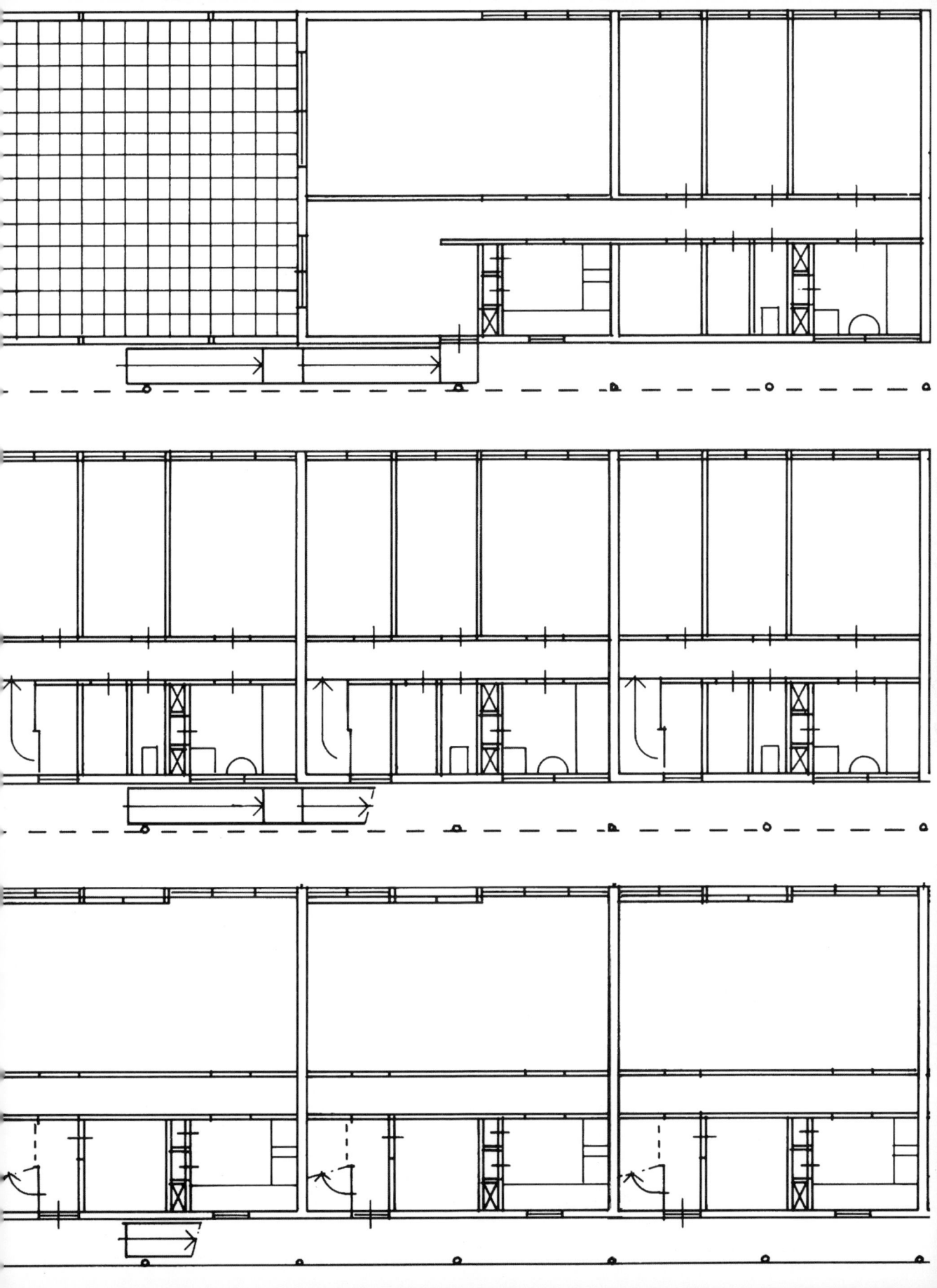

Pendrecht
Neutrality

For the transformation of a zone on the south side of the Rotterdam district of Pendrecht, a densification was achieved within the rhythms and sequences of the original plan by Lotte Stam-Beese from the 1950s. There she used a sequence in construction height of 2, 3, 3.5 and 4.5 storeys. The succession has been amplified by adding two or three storeys on top, achieving greater variety with new series and rhythms of 3, 3.5, 4, 4.5, 6.5 and 7 storeys. There is still no dominant height within this: the neutral order of the original plan is maintained.

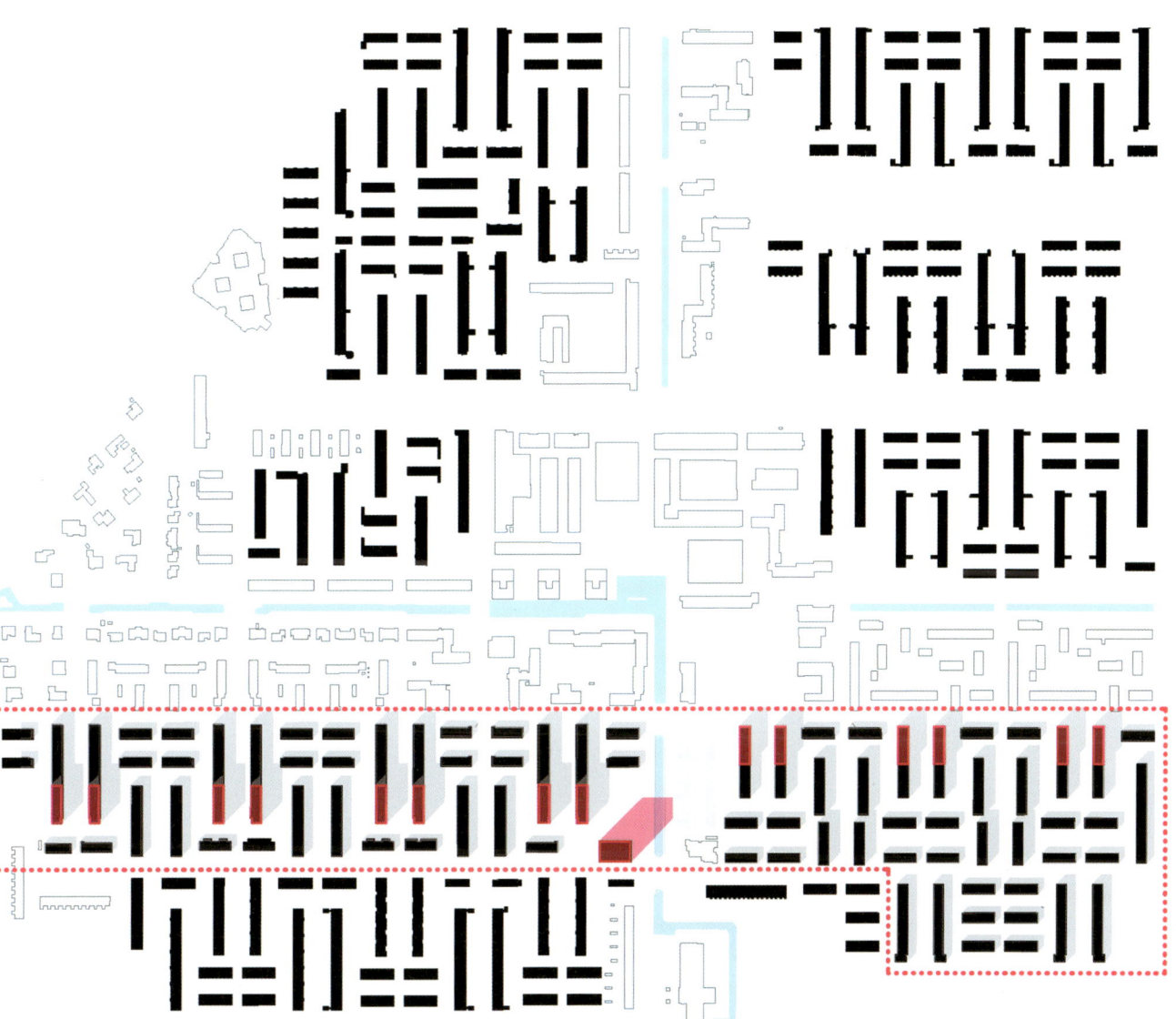

Ouagadougou
Neutrality

Acknowledging and organizing the spontaneous urban growth, an urban plan was devised for an area of the capital of Burkina Faso, by starting from the smallest unit, the dwelling, rather than effecting large–scale interventions. In the traditional house, basic facilities such as sleeping space, a place to cook and storage space are accommodated in a neutral manner within an elementary form, built with clay that is found locally.

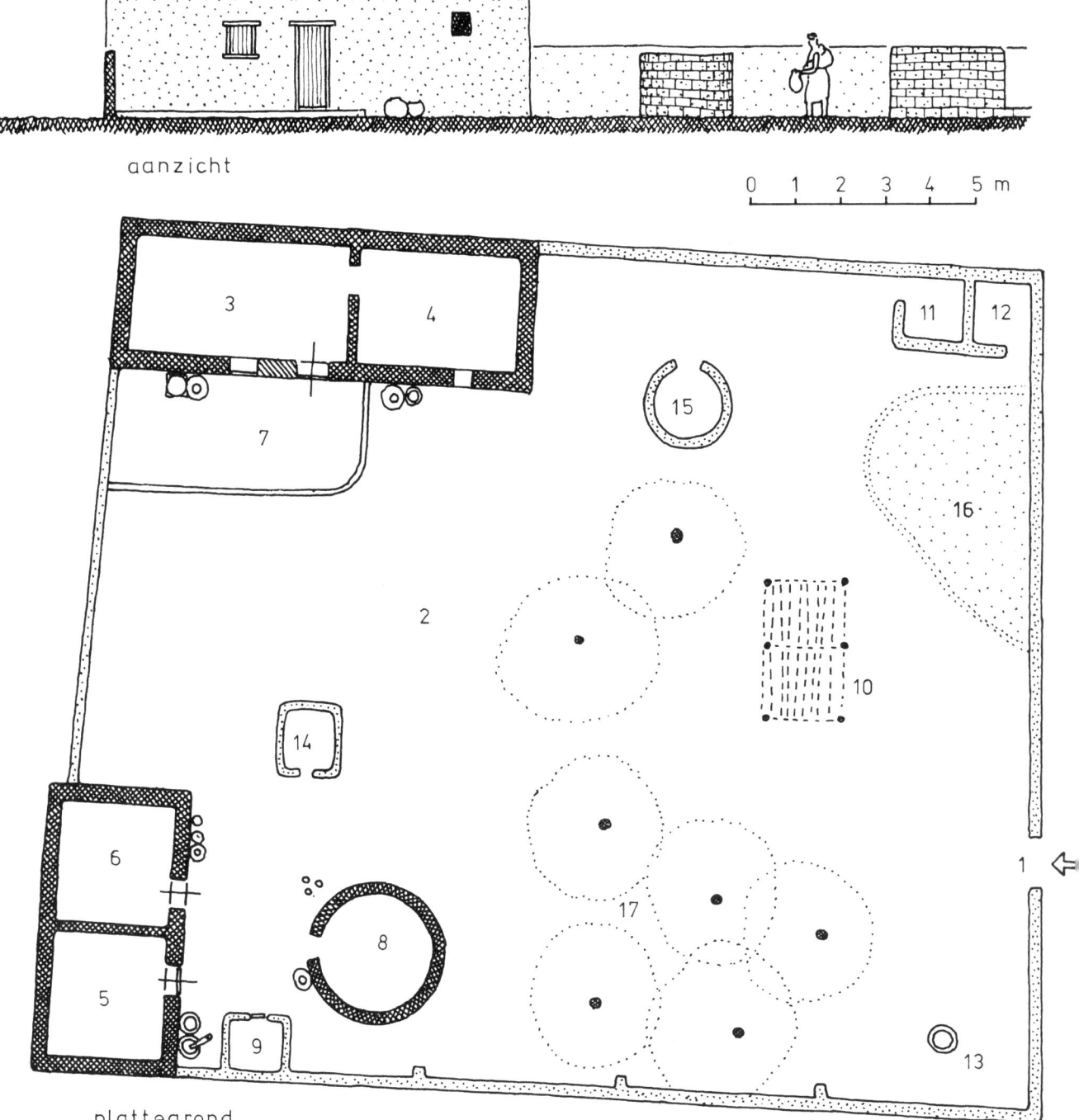

aanzicht

0 1 2 3 4 5 m

plattegrond

LEGENDA

1 ingang

2 erf

woonruimte van:

3 hoofd van de familie

4 vrouw en kinderen

5 getrouwde zoon

6 schoondochter en kinderen

7 verhard terras

8 keuken

9 opslag

10 ontvangst

11 latrine

12 wasplaats

13 afvalput

14 kippenhok

15 geitestal

16 moestuin (inc.)

17 fruitbomen

Erasmusgaarde
Neutrality

The project for the restructuring of Field 17 in the redevelopment of The Hague's Zuidwest district was determined by a number of rules: the basis is a grid of 30 by 30 metres. The ratio of built to non-built land remains as it was in the old situation, but is organized differently. The total floor area of the dwellings is the same, but the construction density was reduced from 90 to 51 units per hectare. • The architecture is organized orthogonally, with horizontals and verticals crossing each other in plan and striking a mutual balance. The project thus adopts the essence of post–war urban planning: buildings in an unbounded space without a hierarchy. The neutrality on an urban planning level is also continued in the facades of the buildings.

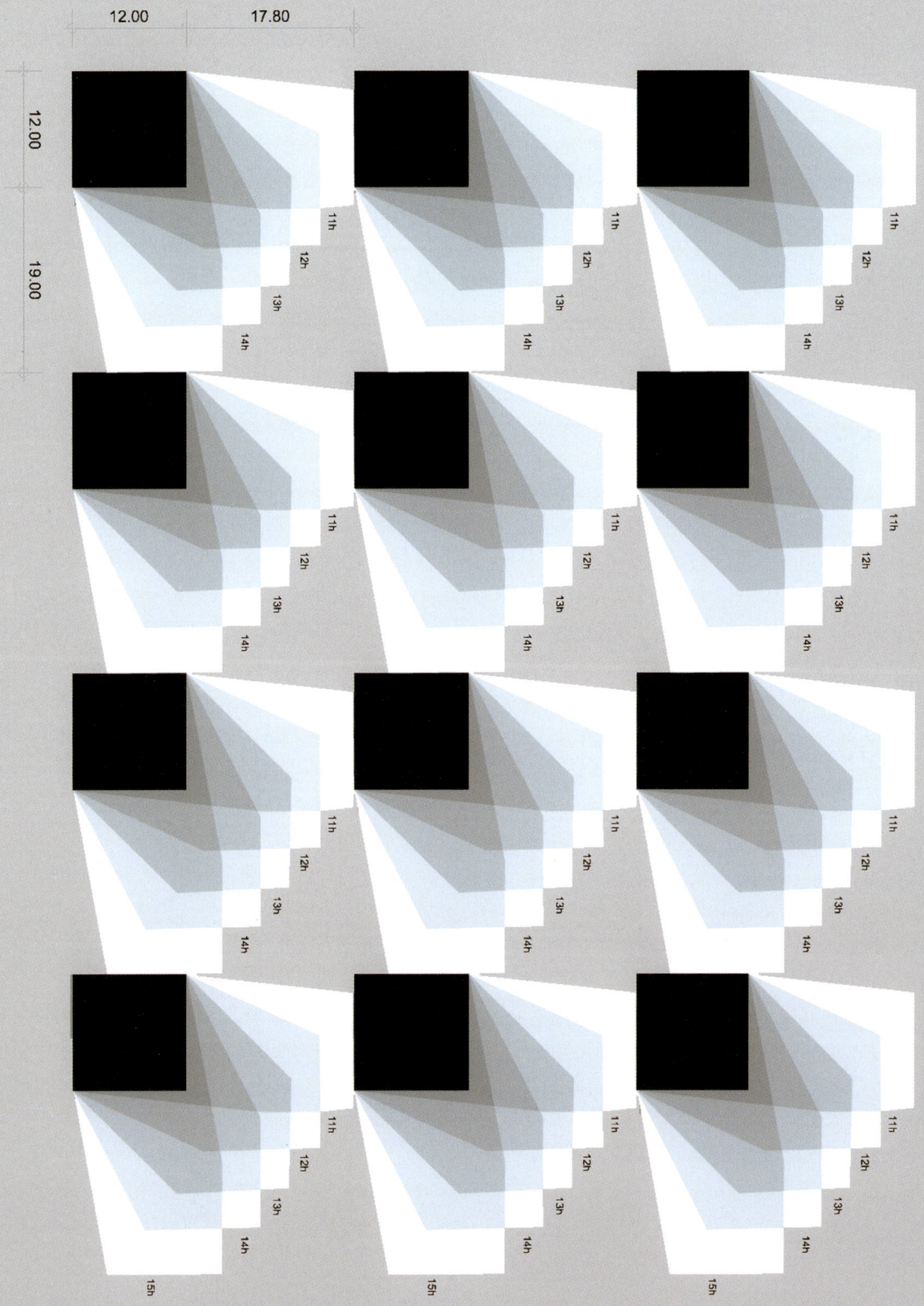
12.00
17.80
12.00
19.00
11h
12h
13h
14h
15h

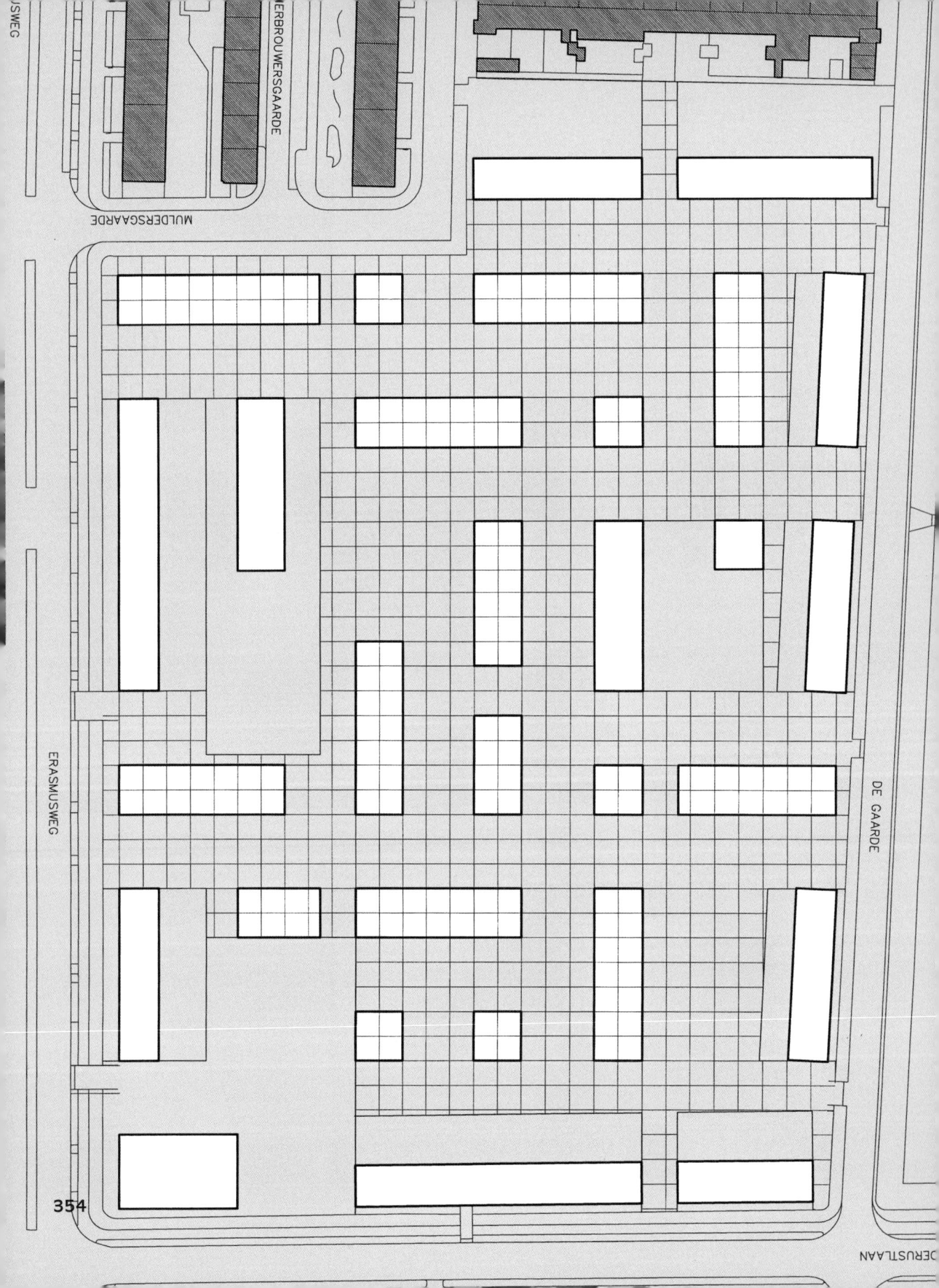

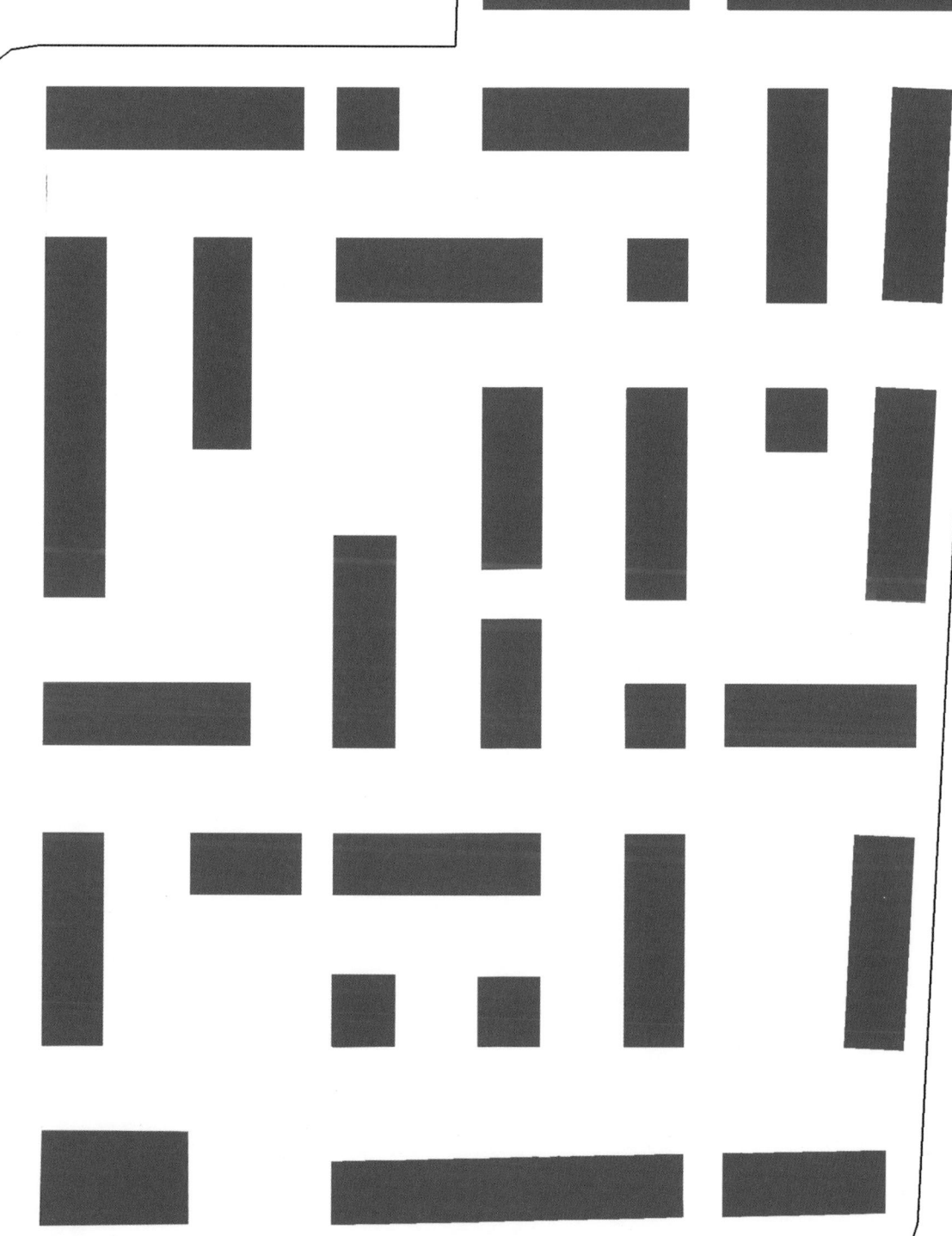

355

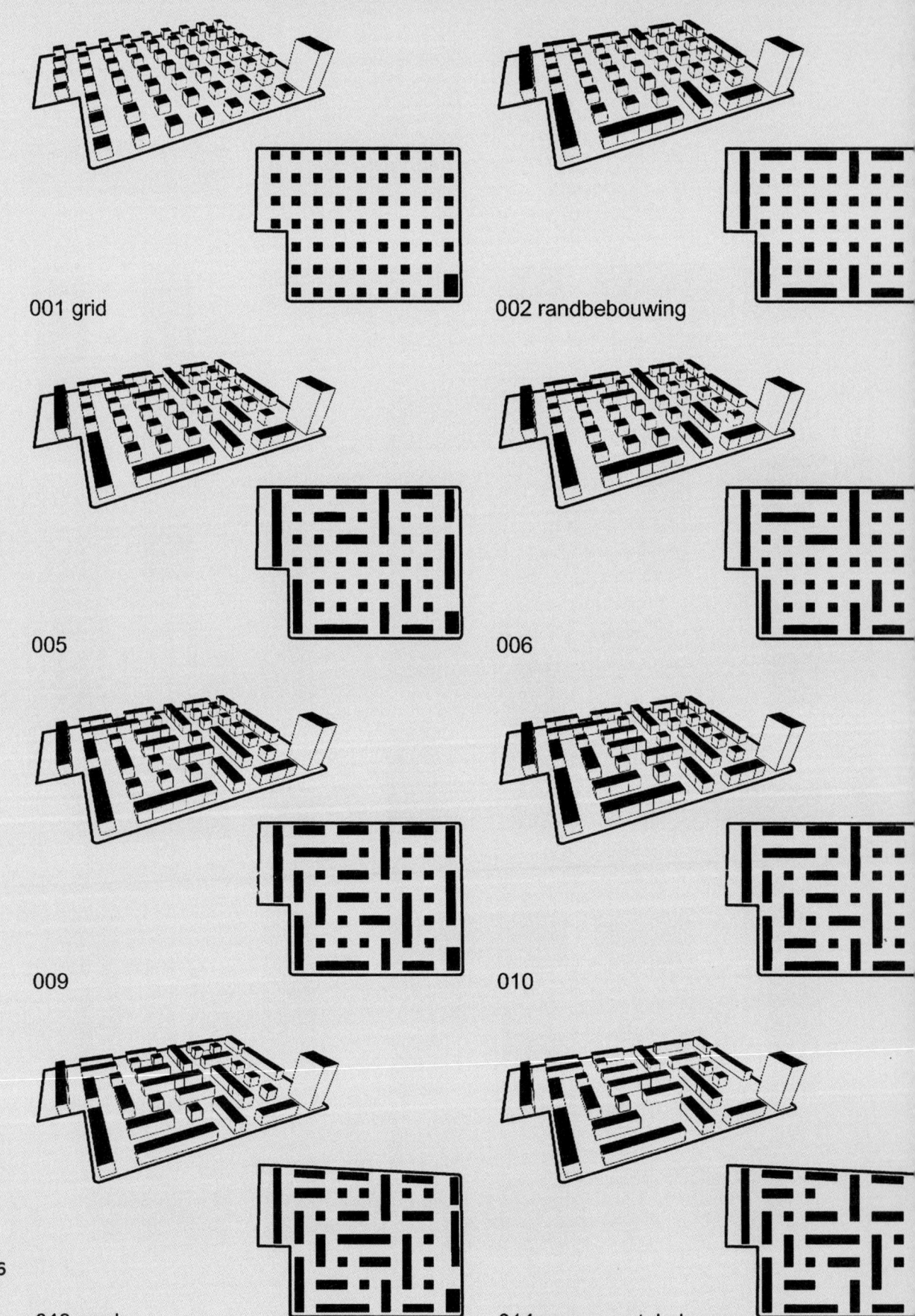

001 grid

002 randbebouwing

005

006

009

010

013 rand

014 monumentale bomen

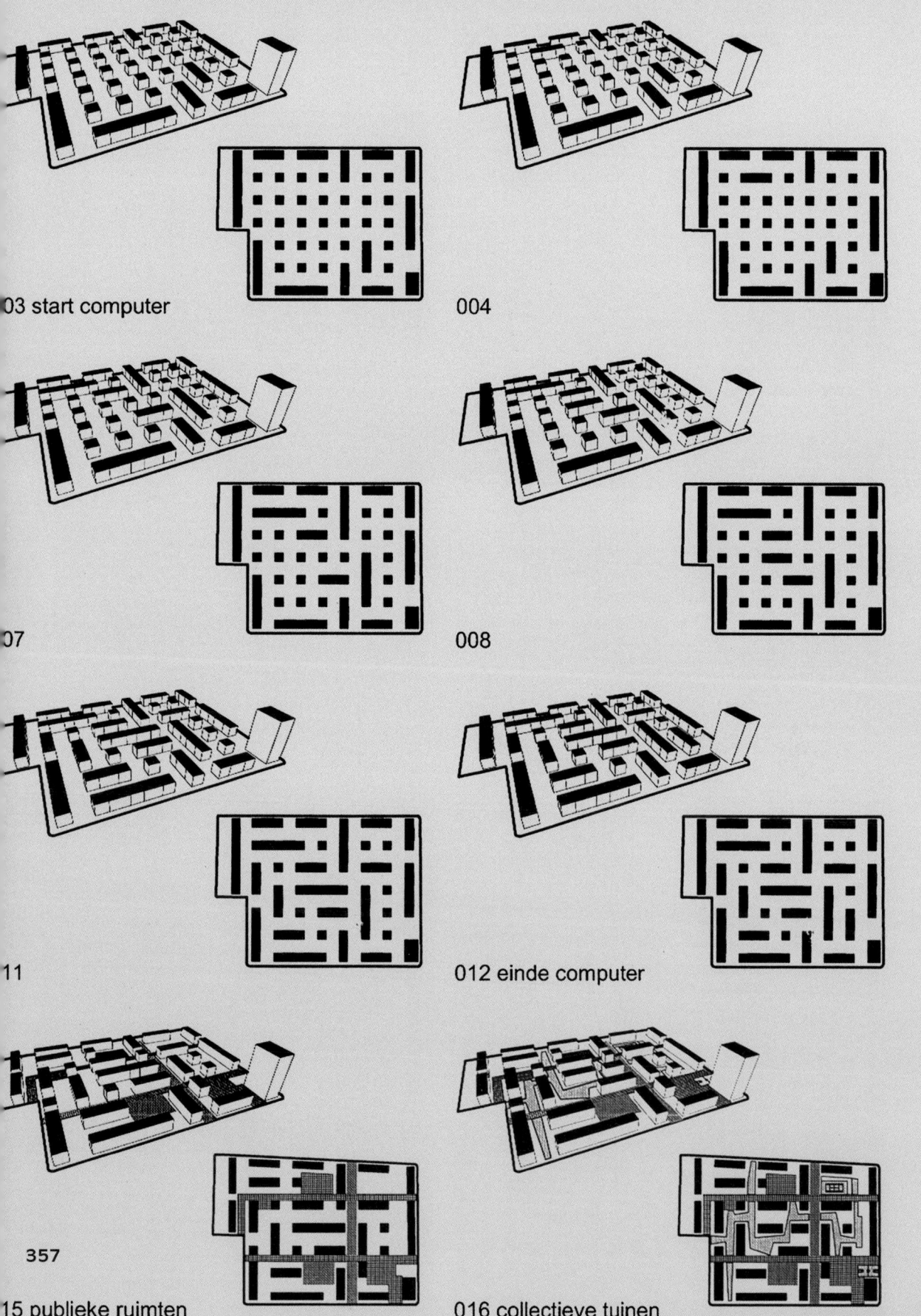

003 start computer

004

007

008

011

012 einde computer

357

015 publieke ruimten

016 collectieve tuinen

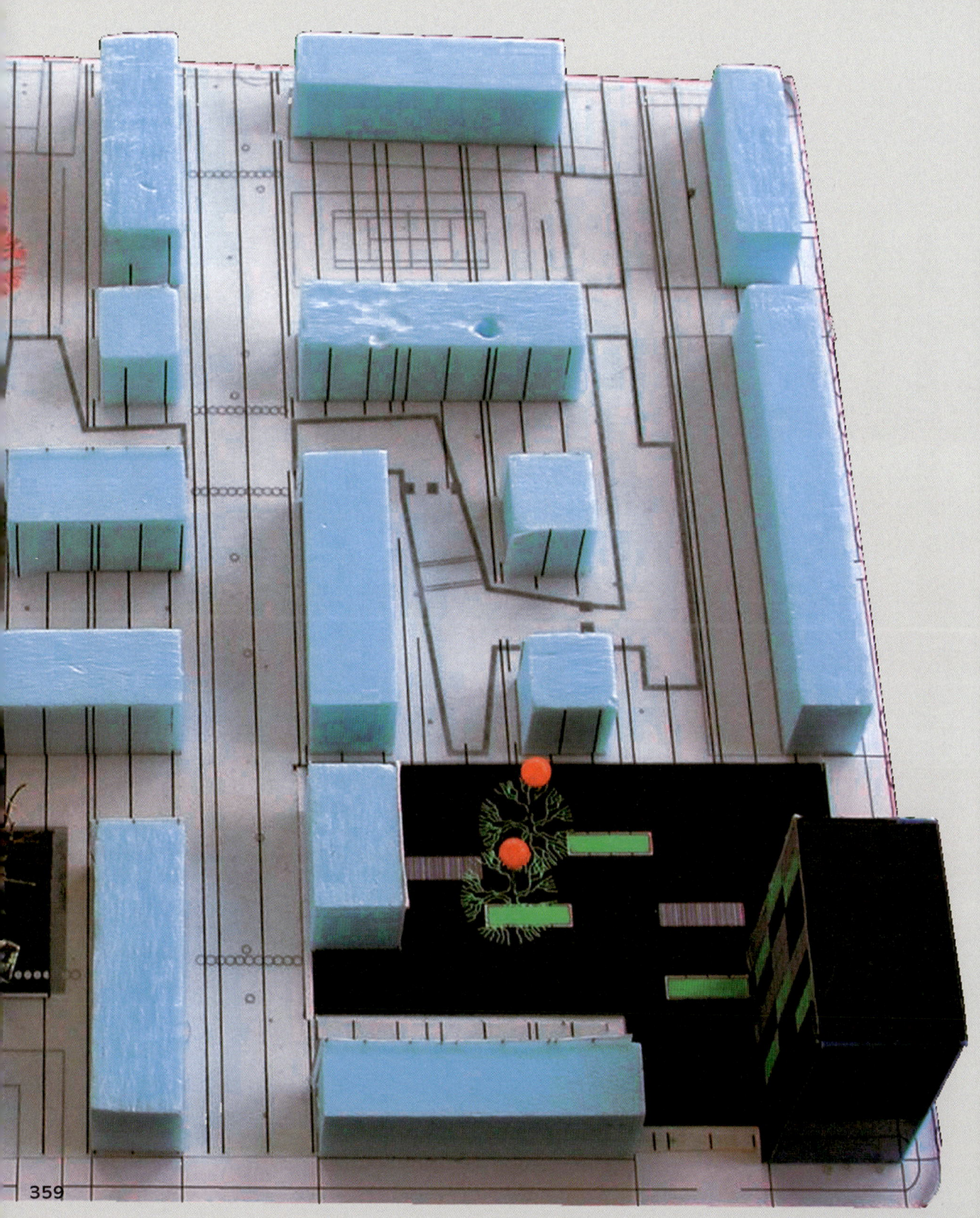

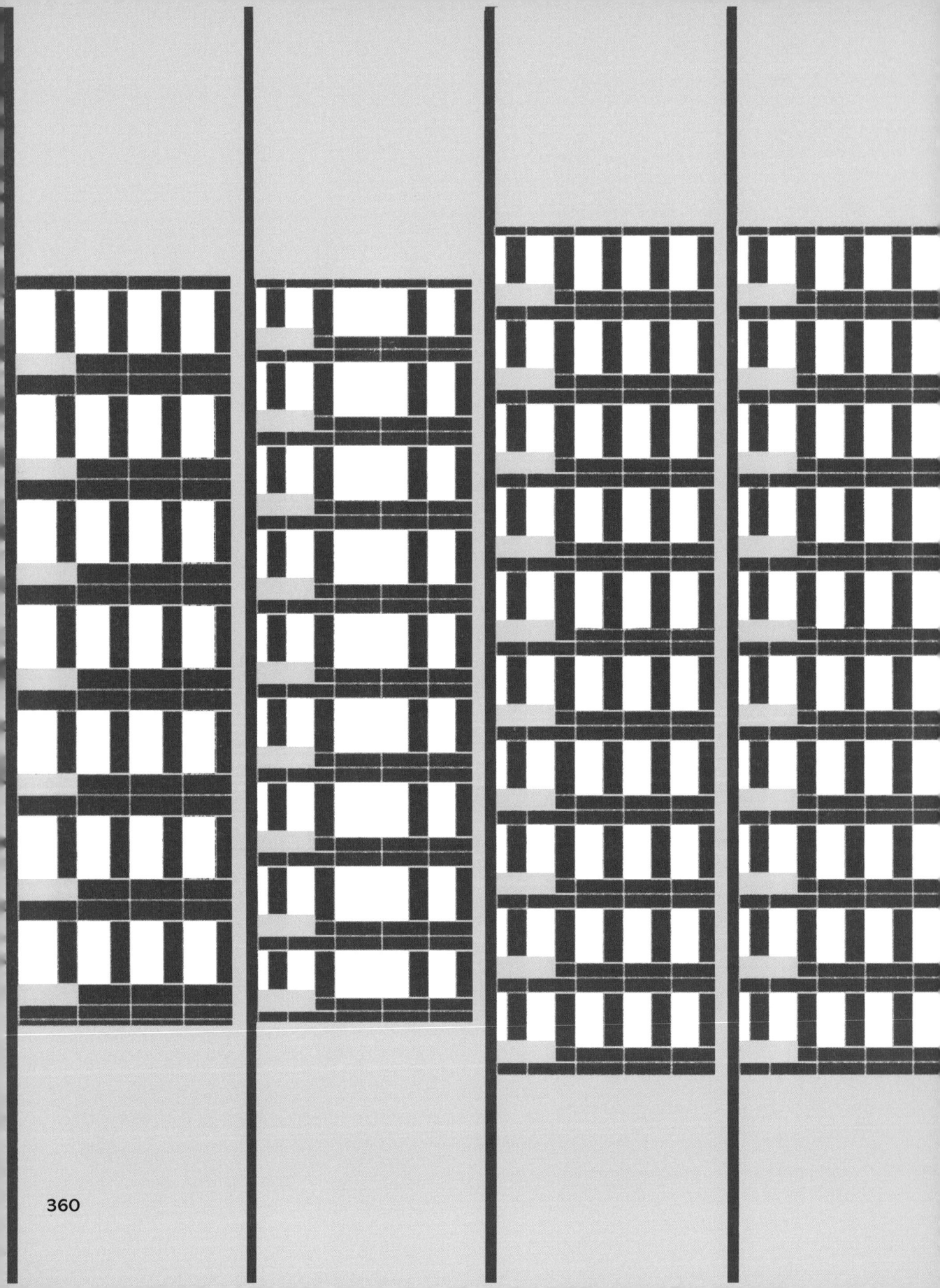

List of Projects

• 600 dwellings, Kairouan, **Tunisia**
Client: The Netherlands National
Committee for Aid to Tunisia, Tunisia
Architect: René van Veen
design 1970/71 – realization
1972/73

• Urban plan Wagadogo Nossin,
Ouagadougou, Burkina Fasso
Client: Vrije Universiteit, Amsterdam
Architect: René van Veen
design 1979/83

• 44 houses Zuiderpolder, **Haarlem**
Client: De Vonk housing association,
Haarlem
design 1986/1987 – realization 1990

• Kunstbeurs RAI (study) Amsterdam
design 1986

• Reprofiling of Frederiksplein
(competition) Amsterdam
design 1987

• Housing for young people
(competition, final round)
design 1987

• 194 dwellings (invited competition)
Geuzenveld–West, Amsterdam
Client: Slokker Vastgoed, Huizen
design 1988

• Floriade entrance zone (competition)
Zoetermeer
design 1988

• 41 dwellings Tanthof, Delft
Client: City of Delft
design 1989

• House annex workspace Zielhorst,
Amersfoort
Client: K. Beyen, Amersfoort
design 1989 - realization 1990

• 238 apartments former AZU site, Utrecht
Client: AmstellandVastgoed, Amsterdam
Project architect: René van Veen
design 1989 – realization 1990

• Expansion of Teylers Museum
(competition, final round) Haarlem
design 1990

• 194 dwellings with parking
Client: Slachthuisterrein (slaughter
house site), The Hague
Principal AWV housing association
design 1989

• Anders Bouwen, Anders Wonen
('Build Differently, Live Differently',
competition) Amsterdam-Noord
design 1991

• Urban masterplan for the Olympic
Stadium site (invited ideas competition)
Client: Bouwfonds Woningbouw nv,
Haarlem in association with
Karthaus/Brans architecten
design 1991

• 70 family house
Prinsenpark-Noord, Rotterdam
Client: Stichting Volkswoningen, Rotterdam
design 1992 – realization 1994

• Renovation of Madurodam,
The Hague (invited competition)
Client: Madurodam Miniature City
Foundation in association with B+B,
stedenbouw en landschapsarchitectuur bv
design 1992

• 72 patio houses **De Aker**, Amsterdam
Client: BPF–Bouw, Amsterdam
design 1993 – realization 1994

• 6 dwellings for mentally handicapped
residents Prinsenpark, Rotterdam
Client: Stichting Voorzieningen voor
Verstandelijk Gehandicapten Rijnmond
(Rijnmond Foundation for Facilities for
the Mentally Handicapped)
design 1993 – realization 1994

• School of the Future (invited ideas
competition)
Client: Twijnstra Gudde, Management
Consultants, Amersfoort
design 1993

• Urban masterplan (invited ideas competition) Czaar Peterbuurt, Amsterdam-Oost
Client: Projektgroep Oostelijke Binnenstad
(Eastern City Centre Project Group)
design 1993

• 96 dwellings Nieuw Sloten, Amsterdam
Client: Smit's Bouwbedrijf, Beverwijk
design 1993 – realization 1995

• 36 dwellings Amstelveenseweg,
Amsterdam
Client: V.O.F. Amstelveenseweg
design 1993 – realization 1996

• 150 dwellings
Prinsenpark-Noord, Rotterdam
(invited competition, 1st prize)
Client: Van Omme & De Groot
design 1993 – realization 1995

• Urban masterplan (invited ideas competition) Wilhelmina Gasthuis site,
Amsterdam
Client: Amsterdam Borough of Oud–West
design 1993

• 4 urban villas
Dedemsvaartweg, The Hague
(coordinating architect)
Client: Zuid–West district housing
association
design 1994 – realization 1995

• Urban masterplan
Chassé site, Breda (invited competition)
Client: Johan Matser, Hilversum
in association with Dirrix Van Wylick
architecten, Breda
design 1994

• Noordkaap housing development
Dapperbuurt, Amsterdam (invited
competition)
Client: De Principaal
design 1994 – realization 1997

• 24 dwellings
Borneo-Sporenburg, Amsterdam, 1st phase
Client: Ontwikkelingsmaatschappij New
Deal
design 1994 – realization 1997

• 42 dwellings
Drielanden, Groningen
Client: Johan Matser, Hilversum
design 1994 – realization 1996

• Econosto site
Feasibility study, Rotterdam
Client: Maasstede Woningontwikkeling
design 1994

• 29 dwellings and 2 shops
Oosterparkbuurt, Amsterdam
Client: De Principaal
design 1994 – realization 1997

- 36 dwellings
Dorp Noord III, Moordrecht
Client: Municipality of Moordrecht
design 1995 – realization 1997

- Laurenskwartier Rotterdam
Manifestatie 50 jaar Wederopbouw -
50 jaar Toekomst ('50 years
Reconstruction – 50 years Future' event)
Client: City of Rotterdam
design 1996

- Ichthus site Laurenskwartier,
Rotterdam (ideas competition)
Client: City of Rotterdam
design 1995

- Urban masterplan Vuosaaris South
Shore, Helsinki (invited ideas competition)
Client: City of Helsinki
design 1996

- 14 urban dwellings
Nonnenveld, Chassé site, Breda
Client: Chassé development
conglomerate
design 1996 – realization 2001

- 38 dwellings
2de **Oosterparkstraat**, Amsterdam
Client: De Principaal
design 1996 – realization 1999

- 30 dwellings
Borneo–**Sporenburg**, Amsterdam,
2nd phase
Client: New Deal
design 1996 – realization 1997

- 24 dwellings
Waterfront, Bergen op Zoom
Client: Amstelland Vastgoed Eindhoven
design 1997 - realization 2000

- 35 dwellings and 11 business units
Borneo-**Sporenburg**, Amsterdam,
3rd phase
Client: New Deal
design 1997 – realization 1999

- Office building **Muiderstraat**,
Amsterdam
Client: Cocon Vastgoed Management
design 1997 – realization 1999

- Architectural study Vreeswijk,
Nieuwegein (invited ideas competition)
Client: Municipality of Nieuwegein
design 1997

- Urban masterplan Westerdokseiland,
Amsterdam (invited ideas competition)
Client: Projectgroep Zuidelijke IJ–oever
(Southern Banks of the IJ Project Group)
design 1997

- 128 dwellings
Leidschendam, Leidschenveen
Client: R.K. WBV Leidschendam
design 1997 – realization 1998/9

- 56 dwellings Hoogvliet, Rotterdam
Client: Van Omme & de Groot bv
design 1997 – realization 1999

- 2 urban villas Dedemsvaartweg,
The Hague (coordinating architect)
Client: Schouten/De Jong
Projectontwikkeling, Rijswijk
design 1998 – realization 2000

- 14 dwellings
Leeuwenhoek cluster, Amsterdam
Client: De Principaal
design 1998 – realization 2004

- Urban masterplan Osdorp,
Amsterdam (invited ideas competition)
Client: Het Oosten housing association,
Amsterdam
design 1998

- 26 dwellings Osseveld–West, Apeldoorn
Client: Ons Huis housing association
Winner of the Municipality of
Apeldoorn Architecture Prize 2003
design 1998 – realization 2003

- 114 dwellings
Leidschendam, Leidschenveen
Client: Woningstichting Vlieterheem
design 1998 – realization 1999

- ±300 dwellings and commercial
amenities Laurenskwartier, Rotterdam
(ideas competition)
Client: ERA Bouw, ABN–AMRO, BAM, IBC
design 1998

- Urban masterplan
Terrasdorp, Amsterdam-Noord
Client: Zomers Buiten, Het Oosten,
AWV housing associations
design 1998

- 50 dwellings
De Waalsprong, Nijmegen, 1st phase
Client: Tweepool Projectontwikkeling, Elst
design 1998 – realization 2000

- 171 dwellings
Meerhoven Noord, Eindhoven
Client: Domein, Woonstichting SWS
design 1998 – realization 2004

- 6 urban **bridges**
IJburg, Amsterdam
Client: Gemeentelijk Grondbedrijf
design 1999 – realization 2002

- 45 dwellings and commercial units
Block **6A**, **Almere** city centre
Client: Blauwhoed / Eurowoningen
design 1999 – realization 2004

- Laurenskwartier Rotterdam
(Urban Development Study)
Client: City Centre section, Rotterdam
Department for Urban Planning and
Housing (dS+V)
design 1999

- 170 dwellings with school and gym-
nasium Katendrecht–Parkkwartier,
Rotterdam (invited competition)
Client: Ballast Nedam Woningbouw,
Rotterdam Department for Urban
Planning and Housing (dS+V)
design 1999

- 116 dwellings and business units
Valkenbos–Zuid, The Hague (invited
competition)
Client: Heijmans Vastgoed, Leiderdorp
design 1999

- 10 dwellings **Het Funen**, Amsterdam
Client: IBC Vastgoed, Woerden
design 1999

- ±30 dwellings Assumburgweg,
The Hague (invited competition)
Client: Schouten/De Jong
Projectontwikkeling, Rijswijk
design 1999

- 43 dwellings
De Waalsprong, Nijmegen, 2nd phase
Client: Portaal Ontwikkeling, Elst
design 1999 – realization 2005

- 34 dwellings **Block 34**,
IJburg, Amsterdam
Client: BouwWerk
design 1999 realization 2005

- Ichthus building, ±150 dwellings
+ commercial units Laurenskwartier,
Rotterdam

Client: Laurenskwartier development conglomerate
design 1999

• 3 residential towers with 168 apartments **Stramanweg**, Amsterdam-Zuidoost
Client: BAM Vastgoed
design 1999 – realization 2005

• 113 dwellings
CiBoGa site, Groningen
Client: IMW collaborative partnership
design 2000

• Urban masterplan and 344 dwellings
Erasmusgaarde, The Hague
Client: HaagWonen housing association
design 2000 – realization 2004

• ±40 houses
Block 26, IJburg, Amsterdam
Client: Eurowoningen, Rotterdam
design 2000

• 40 dwellings
Block 30, IJburg, Amsterdam
Client: Waterstad 3
design 2000 – realization 2004

• Urban masterplan
Westelijk Stationseiland, Amsterdam
(invited ideas competition)
Client: Projectgroep Zuidelijke IJ–oever
(Southern Banks of the IJ Project Group)
design 2000

• **100 x 100**, 64 park villas
Chassé site, Breda
Client: Chassé development conglomerate
design 2000

• Office building Sneller Poort, Woerden
Client: Lingestroom II CV development conglomerate
design 2000

• 110 dwellings
Waterbuurt West, IJburg, Amsterdam
(invited competition)
Client: Smit's Bouwbedrijf, Beverwijk
design 2001

• 63 dwellings
De Bongerd, Amsterdam Noord
Client: De Bongerd development conglomerate
design 2002 – realization 2005

• 72 dwellings
Witbrant-Oost, Tilburg
Client: De Wilde Projectmanagement, Tilburg
design 2002 – realization 2004/5

• 7 **bridges**
Haveneiland-Oost, IJburg, Amsterdam
Client: Gemeentelijk Grondbedrijf (Municipal Real Estate Dept.)
design 2003 – realization 2005

• 78 apartments and commercial facilities
Waterrijk Viermeren, Woerden
Client: Proper Stok, Rotterdam
design 2003 – realization 2005

• 100 x 100, 66 dwellings
(redevelopment) Chassé site, Breda
Client: Chassé development conglomerate
design 2003 – realization 2005

• High-rise tower with 34 apartments
Suytkade, Helmond
Client: Van Wijnen, Waalwijk
design 2003 – realization 2005

• Urban masterplan and 67 dwellings, 1st phase Erasmuspark, The Hague-Zuidwest
Client: Kristal, The Hague
design 2003 – realization 2004/5

• Land division study for the 'Transformation Zone'
Pendrecht, Rotterdam (invited ideas competition, 1st prize)
Client: De Nieuwe Unie / Rotterdam Department for Urban Planning and Housing (dS+V)
design 2003

• 16 dwellings and parking–house
Willibrordusstraat, Amsterdam
Client: EMO Stedelijke Ontwikkeling
design 2003 – realization 2006

• 34 apartments (redevelopment)
Block 34, IJburg, Amsterdam
Client: Bouwwerk
design 2003

• 3 residential towers with 168 apartments (redevelopment)
Stramanweg, Amsterdam–Zuidoost
Client: BAM Vastgoed
design 2003 – realization 2005

• Roof structures
De Waalsprong, Nijmegen
Client: PPO, Nijmegen
design 2004 – realization 2005

• Apartment building Dinteloordstraat, Rotterdam – Pendrecht
Client: De Nieuwe Unie, Rotterdam
design 2004 – realization 2006

• Urban masterplan and 219 dwellings
ROC site, Kanaalzone, Apeldoorn
(invited competition, 1st prize)
Client: Le Clercq Planontwikkeling, Deventer
design 2004 – realization 2006/7

• Residential tower with ca. 60 apartments Erasmuspark, Zuidwest, The Hague
Opdrachtgever: Kristal BV, The Hague
design 2004 – realization 2005/6

• 18 apartments (redevelopment)
Het Funen, Amsterdam
Client: Heijmans IBC Vastgoedontwikkeling, Almere
design 2003 – realization 2004

• Urban masterplan (in association with De Nijl Architecten)
Rotterdam – Pendrecht Transformation Zone
Client: De Nieuwe Unie / Rotterdam Department for Urban Planning and Housing (dS+V)
design 2004 – realization 2006

• Urban masterplan and 291 dwellings, 2nd phase Erasmuspark, The Hague–Zuidwest
Client: Kristal, The Hague
design 2004 – realization 2006/7

Erna van Sambeek

1966–1971 St. Joost Academy for Fine Arts, Breda,
the Netherlands
1972–1983 Amsterdam Academy of Architecture, Amsterdam,
the Netherlands

1984–1985 scholarship from the Ministry of Culture
1986–1987 grant from the Ministry of Culture
1989–1994 Member of the board of the Royal Institute of
Dutch Architects (BNA)
1990 Member of jury, 'The Bronze Beaver', National Prize for
Building and Homes
1991 Gerrit Rietveld Academie, State appointment to the
Department of Architectural Design
1991 Member of jury, Concrete Award
1992–1994 Member of the Amsterdam Council for Monuments (ARM)
1994 Member of jury, Encouragement Award for the Arts,
Amsterdam
1994–1995 Project supervisor Amsterdam School of the Arts
(AHK), Amsterdam Academy of Architecture
1994–1995 Member of jury, Archiprix
1995 Member of the Assessment Committee for Architecture
Education for the Architects Register (SBA)
1995 Member of jury, Prix de Rome
1995 Member of jury, Young Architects Initiative, Eindhoven
1995 Member of jury, New Country Estates Competition
1995 Member of jury, Apeldoorn Architecture Prize
1996–2000 Member of the selection committee for Individual
Subsidies for Architecture of the Netherlands Foundation for
Visual Arts, Design and Architecture (Fonds BKVB)
1997 Member of jury, Bruges Concert Hall International
Architecture Competition
1997 Member of jury, Competition for the seat of the Provincial
Government of Flemish Brabant
1998 Delft University of Technology (TU Delft), Lecture at the
Faculty of Architecture
1998 University of Leuven (KU Leuven), Lecture at the Faculty of
Architecture
1999 Member of jury, Architecture competition for the Arcade
building, Apeldoorn
2000–2003 Member of the board of the Fonds BKVB

Since 1986 associated with René van Veen, established bureau
under the name Van Sambeek & Van Veen, Architecten BNA.

René van Veen

1957–1960 Higher Technical School (HTS, now Rotterdam
University), Rotterdam
1961–1965 Royal Academy of Art (KABK), The Hague
1965–1967, 1974–1978 Amsterdam Academy of Architecture,
Amsterdam

1968 Kahn and Jacobs architecture bureau, New York, U.S.A.
1969 Aldo van Eyck archictecture bureau, Amsterdam
1970 Travaux du Midi, Marseilles, France
1970–1973 independent architect for The Netherlands National
Committee for Aid to Tunisia
1974–1978 Van Eyck en Bosch architecture bureau, Amsterdam
1979–1983 Neighbourhood restructuring, Ouagadougou, Upper
Volta (now Burkina Faso)
in association with the Vrije Universiteit, Amsterdam
1980–1988 supervisor for graduation projects, Faculty of
Architecture, Eindhoven University of Technology
since **2003** Member of the Amsterdam Commission for Aesthetics
and Monuments

Since 1986 associated with Erna van Sambeek, established bureau
under the name Van Sambeek & Van Veen, Architecten BNA.

www.vsvv.nl

Collaborators

Rita Abreu
Victor Ackerman
Lennart Aertsen
Joao Aguiar
Jacqueline Alling
Gijs Baks
Carin Barreveld
Yvonne Beck
Sanne Beeren
Meta Berghauser Pont
Peter Boer
Martien de Boer
Debby Borghouts
Sjoerd van den Bos
Peter Brouwer
Steven Brunsmann
Maria Ana Caldas
Cyrus Clark
Claire Crowley
Peter Crowley
Gabrielle Demme
Ofri Earon
Annemarie Eijkelenboom
Michel van Erk
Paul van der Erve
Eva Geering
Casper le Fèvre
Romy Franke
Renée van Gennip
Mike Gieskens
Kirsten Hamstra
Per André Haupt
Lenny Heijboer
Rob Hendriks
Hendrine den Hengst
Marian Hogezand
Quintus Huber
Martin IJtsma
Arjenne Jetzes
Eddy Joaquim
Gitte Johannesen
Stefan van de Kar
Gülsen Karatas
Joke Kaufmann
Renato Kindt
Peter Knaven
Melinda Koopman
Mike Korth
Esmeralda Korver
Johan Krol
Lia Kroon-Uitermark

Tanja van der Laan
Thorsten Lang
Maarten Laoût
Petra Lieke
Stijnie Lohof
Uta Lorenz
Marc Losenegger
Jannegien Luursema
Evangelos Lykos
Luc Malnati
Ralph van Mameren
Agnes Mandeville
Bob Mantel
Pedro Nuno Martins
Stani Michiels
Stuart Milne
Floor Moormann
Bas Morsch
Mireia Mosset
Koen Mulder
Anne-Valerie Nahrath
Pepijn Nolet
Michael Noordam
Gerbrand van Oostveen
Daniël Peters
Madeleine Peters
Pelle Poiesz
Uma Poskovic
Prashant Pradhan
Jef van den Putte
Oliver Rasche
Ori Rozental
Paul Salomons
Marcel Scherrer
Tatjana Schneider
Guido Schot
Jasper Schweigman
Wim Sjerps
Nicolette Sloots
Nuno Sousa da Silva
Bart Spanjer
Esther Stevelink
Wilbert Swinkels
Björn Utpott
Jaap Veerman
Floor Visser
Bart van der Vossen
Yuri Werner
Letitia Williams
Martin Young
Gitta Zaeschke

Bibliography

René van Veen, *3 years of building in Tunisia*, Forum, vol. 25, nos. 5-6 (1976)

René van Veen, *Ruimtelijke Ordening in Ouagadougou*, Plan, vol. 11, May 1980

René van Veen, *Physical planning in Ouagadougou, Upper Volta*, Open House, vol. 5, no. 4, 1980

René van Veen, *Drie steden in Portugal*, research report, Eindhoven University of Technology (TUE), Eindhoven, 1983

René van Veen, *De illegale bouw in Lissabon*, Wonen TA/BK, nos. 22-23, November 1983

Wim J. van Heuvel, *Experimentele woningen voor de Zuiderpolder*, AB. Architectuur/Bouwen, no. 2, 1987

Cees Swinkels, *Haarlems woningwet-experiment*, De Architect, March 1987

Various authors, *Experimentele woningbouw Zuiderpolder Haarlem*, Items, special edition, 1987

Sophie Rousseau, *Individualistes, mais d'un commun accord*, L'Architecture d'Aujourd'hui, no. 257, June 1988

Various authors, *Teylers Museum, plannen voor uitbreiding*, 1990

Erna van Sambeek and René van Veen, *Geen rij huizen, maar één gebouw*, AB. Architectuur/Bouwen, no. 2, 1991

Mariëtte van Stralen and Bart Lootsma, *Autonomie en context. Woningbouw van Erna van Sambeek*, De Architect, April 1991

Hans Ibelings, *Twee Woongebouwen / Two Residential Blocks*, Jaarboek Architectuur in Nederland 1990–1991 / Architecture in the Netherlands. Yearbook 1990-1991. Rotterdam: 1991

Jos van Eldonk, *Erna van Sambeek, Architectuur der vanzelf-sprekendheid*, Archis, June 1991

Tom Maas, *Berlages Amsterdam-Zuid voltooid*, AB. Architectuur/Bouwen, no. 12, 1991

Han Slavik, *Wohnungsbau in den Niederlanden*, AW. Architektur + Wettbewerbe, no. 147, 1991

Overzichtspublicatie 1991 (1991 overview), Rijksprijs voor Bouwen en Wonen *De Bronzen Bever* (National Prize for Building and Homes *The Bronze Beaver*)

John Cüsters, *Overmaat nodig voor flexibiliteit*, Bouw, no. 7, April 10, 1992

Rue Solidarnosc à Haarlem, in L'Architecture d'Aujourd'hui, no. 282, September 1992

Sozialer Wohnungsbau in Haarlem, NL, Detail, series 1994.1

Timo de Grefte et al., *De school van de toekomst*, Tracy Metz (ed.), Amersfoort, 1994

Tom Maas, *Vindt de stad uit!*, AB. Architectuur/Bouwen, June–July 1994

Margriet Pflug and Marc A. Visser, *Stedenbouw en Kleur*, Bussum, 1995

Liesbeth Melis and Jos Roodbol, *Opkomst van de patiowoning in stedelijke contex'*, De Architect, May 1995

Ole Bouman (ed.), *Rotterdam 2045. Visies op de toekomst van stad, haven en regio*, publication to accompany the

Manifestatie Rotterdam 50 jaar wederopbouw - 50 jaar toe-komst ('50 years Reconstruction – 50 years Future' event), Rotterdam, 1995

Liesbeth Melis, *Zoeken naar de ruimte om de dingen*, De Architect, February 1996

Woningbouw AmsterdamOsdorp in Jaarboek Architectuur in Nederland 1995–1996 / The Architecture Annual 1995–1996, Rotterdam, 1996

Sichtmauerwerk im Wohnungsbau Drei Beispiele aus den Niederlanden, Detail, serie 1996, 7

Bart de Vries, *Het baksteendetail herontdekt*, in Baksteen, September 1996

Y. Sens (ed.), *In vorm, Projecten en bijbehorende kosten 1994/1995*, The Hague, 1996

Arjen Oosterman, *Woningbouw in Nederland. Voorbeeldige architectuur van de jaren negentig / Housing in the Netherlands: Exemplary architecture of the Nineties*, Rotterdam,1996

Friederike Schneider (ed.), Grundrißatlas: *Wohnungsbau / Floor Plan Atlas: Housing*, second edition, Basel, 1997

Marieke van Zalingen, *Erna van Sambeek: Kijk en luister naar de omgeving*, in Eigen Huis & Interieur, January 1997

Arjen Oosterman, *Positions. Van Sambeek & Van Veen*, Archis, no. 2, February 1997

Verslag Woonstichting De Key, annual report, Amsterdam: 1998

John Westrik, *The campus model as magic formula*, Archis, no. 7, 1998

Carla Debets, *Niet één plek hetzelfde. Staalkaart van Stedelijke Interventies in Amsterdam*, Bouwwereld, no. 13, 1998

Walter Stamm–Teske, *Preiswerter Wohnungsbau in den Niederlanden 1993-1998*, Verlag Bau und Technik, Düsseldorf, 1998

Karel Vandenhende, Erna van Sambeek and René van Veen, *Erna van Sambeek*, Konstruktief, no. 33, 1999

Arian Mostaedi, *Residential Complexes*, Barcelona, 1999

Brigitte de Maar, *Een Zee van Huizen. De woningen van New Deal op Borneo/Sporenburg*, Bussum, 1999

Maarten Kloos and Marlies Buurman (ed.), *Amsterdam Architecture 1997–1999*, Amsterdam, 2000

Gerald van der Kaap, *Wonen Forever, De Principaal BV anno 2000*, Amsterdam, 2000

Hans Ibelings (ed.), *The Artificial Landscape*, Rotterdam, 2000

Rob van Gool, Lars Herstelt, Frank Bertolt Raith and Leonard Schenk, *Das Niederländische Reihenhaus, Serie und Vielfalt*, Stuttgart/Munich, 2000

Maarten Kloos and Dave Wendt (ed.), *Formats for Living. Contemporary floor plan in Amsterdam*, Amsterdam, 2001

Borneo/Sporenburg Projects, Amsterdam, the Netherlands, AU, no. 380, May 2002

Winfried Janssen, *Experimenteel bouwen in Zuidwest*, Den Haag Zuidwest, een naoorlogs stadsdeel in verandering, Bussum, 2002

Jacqueline Tellinga, *De Grote Verbouwing*, Rotterdam, 2004

Picture credits

Aldershoff, Roos: 97, 153
BN / De Stem, Johan van Gurp: 99
Bosch Slabbers: 76
Dienst ROEZ Gemeente Groningen / Aviodrome Luchtfotografie, Lelystad: 191
Gemeente Amsterdam / Aviodrome Luchtfotografie, Lelystad: 90-91
Gemeente Den Haag / Aviodrome Luchtfotografie, Lelystad: 103, 171
Kramer, Luuk: 16, 17
Meer, Hans van der / Hollandse Hoogte: 39-40
Musch, Jeroen: 2-3, 55, 58-59, 116-117, 123, 124
Odé, Johannes: 329-330
Richters, Christian: 52-53, 100-101, 165, 166, 167, 310-311
Ruig, Peter de: 172
Schutte, Gert: 66-67
Taudin Chabot, Raymond: 154, 155, 188-189
Terlinden, Christophe (photo Blaise Adilon): 119
Torckler, Darryl / Getty Images: 297-298
Van Sambeek & Van Veen: 61, 83, 95, 104, 105, 146, 147, 199, 209, 247, 257, 302, 303, 305, 317, 341, 346-347, 358-359
Veen, René van: 13, 25, 27, 73, 74-75, 162, 163, 178-179, 185, 186, 195, 201, 202, 211-212, 215, 224, 225, 226, 227, 241, 244-245, 286-287, 349, 351
Vlugt, Ger van der: 78, 79, 80-81, 114-115, 137, 138-139, 279
Werlemann, Hans: 31, 258, 259, 260, 261, 276-277
Wessing, Koen: 181-182

Scale models: Van Sambeek & Van Veen: 31, 61, 83, 95, 104, 105, 146, 147, 202, 209, 247, 276-277, 286-287, 302, 303, 305, 317, 341, 346-347, 358-359; Spaaij bv, Amersfoort: 178-179; DRO vorm: 258, 259, 260, 261

Drawings and renderings: Van Sambeek & Van Veen, except for: 7 arts visuals, Rijen; artist impression commissioned by De Wilde Projectmanagement: 35, 197

Acknowledgements

Design images — Van Sambeek & Van Veen Architecten, Amsterdam
with thanks to Agnes Mandeville
Typography — Rudo Hartman, The Haque
Production — Caroline Gautier, NAi Publishers
Translation — Andrew May, Amsterdam
Copy editing — Margreet Udo, Zuiderwoude
Printing — Drukkerij Die Keure, Bruges
Paper — Arctic Volume 130 g.
Publisher — Simon Franke, NAi Publishers

This publication was made possible, in part,
by the Netherlands Architecture Fund.

NAi Publishers is an internationally orientated publisher specialized in devel-
oping, producing and distributing books on architecture, visual arts and related disciplines.
www.naipublishers.nl

Available in North, South and Central America through D.A.P./Distributed Art
Publishers Inc., 155 Sixth Avenue 2nd Floor, New York, NY 10013-1507, tel
+1 212 627 1999, fax +1 212 627 9484, dap@dapinc.com

Available in the United Kingdom and Ireland through Art Data, 12 Bell
Industrial Estate, 50 Cunnington Street, London W4 5HB, tel +44 208 747
1061, fax +44 208 742 2319, orders@artdata.co.uk

ISBN 90-5662-365-6
Printed and bound in Belgium